T0172110

Women in Engineering and Science

Series Editor

Jill S. Tietjen
Greenwood Village, CO, USA

The Springer Women in Engineering and Science series highlights women's accomplishments in these critical fields. The foundational volume in the series provides a broad overview of women's multi-faceted contributions to engineering over the last century. Each subsequent volume is dedicated to illuminating women's research and achievements in key, targeted areas of contemporary engineering and science endeavors. The goal for the series is to raise awareness of the pivotal work women are undertaking in areas of keen importance to our global community.

More information about this series at http://www.springer.com/series/15424

Deborah Jean O'Bannon

Editor

Women in Water Quality

Investigations by Prominent Female
Engineers

 Springer

Editor
Deborah Jean O'Bannon
Department of Civil & Mechanical Engineering
University of Missouri–Kansas City
Kansas City, MO, USA

ISSN 2509-6427   ISSN 2509-6435   (electronic)
Women in Engineering and Science
ISBN 978-3-030-17821-5   ISBN 978-3-030-17819-2   (eBook)
https://doi.org/10.1007/978-3-030-17819-2

This Springer imprint is published by the registered company Springer Nature Switzerland AG
The registered company address is: Gewerbestrasse 11, 6330 Cham, Switzerland

Foreword

Why would we benefit from a book to Women in Water Quality? One of the answers lies in the benefits of diversity. Having more than one gender contribute to science and engineering allows for the expansion of viewpoints and foci. This is quite evident in the text as the reader will discover that the addition of women to this field helped launch what one might have formerly called sanitary engineering into environmental engineering. Many of these springboard efforts are documented in the chapter on the History Pioneering Women in Water Quality, spanning from the early 1870s to famous women scientists in the 1900s. And then, the text expands to state-of-the-science chapters covering water quality issues and solutions in both engineered and natural systems.

The chapters on water quality in engineered systems provide a wealth of information for scholars and practitioners to use in their work both in the United States and abroad. These include emerging technologies that allow for a more sustainable future such as combining engineering and nature-based solutions to stormwater management, reuse of non-potable water, optimizing bio-based remediation or biofilms, and novel nutrient separation techniques. The authors bring a global perspective to these issues also adding international solutions for water quality and prevention of viral outbreaks. The final chapters focus on modeling and fate of various contaminants in natural systems. Understanding how these various chemicals migrate and react in the natural environment is crucial to being able to develop effective prevention or remediation techniques. I am honored to be one among the many women who have dedicated their scholarly endeavors to ensuring improvements in water quality at home and abroad.

Civil and Environmental Engineering
Lamar University
Beaumont, TX, USA

Liv Haselbach, PE, PhD, F.ASCE, LEED AP BD+C

Preface

Women in Water Quality: Investigations by Prominent Female Engineers is a collection of scientific papers by 12 women scientists and engineers working in the area of water quality. The scientific papers are prefaced by our rich history – the women who came before us in water quality, starting with the first female graduate from the Massachusetts Institute of Technology – Ellen Swallow Richards. Brief biographies of Ruth Patrick, Rachel Carson, and Sylvia Earle follow in chronological order. Ellen Richard's 1873 chemistry thesis is reprinted in the book's appendix.

The contemporary and cutting edge research of the book's technical authors is divided into two major sections: water quality in engineered systems and water quality in natural systems. Each technical paper is presented with the author's biography.

It has been a privilege to learn about the authors' ingenuity, technical expertise, and intrinsic passion for water quality.

Kansas City, MO, USA   Deborah Jean O'Bannon

Contents

About the Editor

Deborah Jean O'Bannon is a Professor Emerita of Civil Engineering at the University of Missouri-Kansas City (UMKC). She is also a Professional Engineer, with registration in Iowa, Missouri, and Kansas. She received her B.S. in civil engineering from the Massachusetts Institute of Technology with emphasis in environmental engineering and learned about Ellen Swallow Richards (see Chap. 1). She then went to work for the Massachusetts Department of Environmental Quality Engineering (DEQE) (now the Department of Natural Resources), where she designed water quality surveys and conducted water quality modeling and wasteload allocations for the Housatonic River, Assabet and Concord Rivers, the North Shore coastal area, and Boston Harbor. She received her M.Eng. from Manhattan College, where she worked with Dr. Donald O'Connor and Dr. Robert Thomann. She returned briefly to the DEQE, before moving to the University of Iowa to work on her Ph.D. with Dr. Jerry Schnoor and Dr. Forrest Holly. Her doctoral research was on the movement of soluble pollutants in a run-of-the-river impoundment of the Iowa River and was supported in part by a Dissertation Fellowship from the American Association of University Women. She completed dye field studies and a 2-D model.

Dr. O'Bannon obtained a faculty position at the University of Missouri in 1989. She has had research grants and publications in riverine modeling, immunoassay testing of aquatic pollutants, and urban rain garden systems. She also published and was supported in the areas of undergraduate pedagogy and engineering leadership. She routinely taught fluid mechanics, undergraduate water and wastewater treatment design, leadership and ethics, water quality modeling, limnology, computational hydraulics, design of experiments, and civil engineering capstone design. Dr. O'Bannon redesigned the capstone class in 2003 to accommodate real-world projects for industry clients to give the students a design studio that would prepare them for professional practice. Dr. O'Bannon and UMKC received an award for connecting professional practice and education from the National Council of Engineering Examiners in 2009 (the inaugural year of the award).

Dr. O'Bannon has been very active within the Society of Women Engineers (SWE), becoming a SWE Fellow in 2002 and was named the 2017 Distinguished Engineering Educator by SWE. She has served on the national board of directors, the *SWE* magazine Editorial Board, and is active within the local Kansas City section. She is also active within the American Society of Civil Engineers (ASCE), where she is also an ASCE Fellow and serves on the editorial board of the *Journal of Professional Issues in Engineering Education and Practice*.

Dr. O'Bannon is active in Kansas City affairs, having served for 5 years on the Kansas City Wet-Weather Community Panel, and on Kansas City's Ethnic Enrichment Commission since 2006. She hosted a weekly talk show about civil engineering on KCMO TalkRadio 710 for 2+ years, titled "Water, Water Everywhere" and then rebranded the show as "Building Kansas City." The radio show is archived in UMKC's Marr Sound Archive. She is a passionate leadership coach: she trained women faculty for leadership positions under a 2003 NSF grant, regularly taught/moderated a graduate-level class on engineering leadership, lead a national mentoring program with women engineering faculty through SWE, and is recognized internationally as an academic leadership role model.

Dr. O'Bannon grew up around water in the Croton Reservoir district of the New York City water supply network, has canoed in the Saranac Lake system and Delaware River, taught swimming for a number of years, and was a Red Cross Water Safety Instructor at one point. Throughout her life, she has been drawn to water and enjoys the intersection of aquatic issues and mathematics.

Part I
History

Chapter 1
Pioneering Women in Water Quality

Jill S. Tietjen

Abstract Women's contributions in water quality have revolutionized the thinking of society about how to interact with the Earth's natural resources. This chapter covers four prominent women pioneers in water quality. Ellen Henrietta Swallow Richards, designated by *Engineering News Record* as "the first female environmental engineer," developed the first state water quality standards in the late 1800s. Ruth Patrick, after whom the biodiversity tenet the Patrick Principle is named, matched the types and numbers of diatoms in water to the type and extent of water pollution and invented the diatometer to collect and measure those diatoms. Rachel Carson, credited as the catalyst for the environmental movement of the 1960s and 1970s that continues today, wrote extensively about the oceans in addition to authoring *Silent Spring*, an exposé on pesticides. "Her Deepness" oceanographer Sylvia Earle is working today to preserve the world's oceans.

1.1 Introduction

Women's contributions in water quality have revolutionized the thinking of society about how to interact with the Earth's natural resources. The first water purity tables were developed by a woman. One of the key measures of water pollution in freshwater bodies was developed by a woman. A woman is credited with sparking the environmental consciousness of the 1960s and 1970s that led to Earth Day and today's environmental movement. A woman is taking the lead to save our oceans. The women briefly described in this chapter framed the science and ethos of the chapter authors that follow in this volume.

J. S. Tietjen (✉)
Technically Speaking, Inc., Greenwood Village, CO, USA

© Springer Nature Switzerland AG 2020
D. J. O'Bannon (ed.), *Women in Water Quality*, Women in Engineering and Science, https://doi.org/10.1007/978-3-030-17819-2_1

1.2   Ellen Henrietta Swallow Richards (1842–1911)

Ellen Henrietta Swallow Richards has been honored by *Engineering News Record* as one of the top environmental engineering leaders of the last 125 years. She was a true Renaissance woman and a pioneer who opened up many doors for the scientific education of women. It was in 1999, when the magazine also designated her as "the first female environmental engineer." Her immense legacy includes seminal contributions in the areas of environmental and sanitary engineering, groundbreaking contributions toward water and air purification, leading edge analysis of food and human diet, and the design of healthier and safe buildings. Richards produced the first state water purity tables in 1887, helped establish the first systematic course in sanitary engineering at the Massachusetts Institute of Technology (MIT), founded the science of home economics, is called the "mother of ecology," and was one of the founders of what today is called the American Association of University Women [1–3]. As with most women of this era, her beginnings were not quite as auspicious as these many accolades might indicate.

Richards (neé Swallow) was born in 1842 to parents who were both school teachers and she was encouraged to pursue the best education possible with her family's limited resources: much of her schooling was at home. As with many young women of the time, her education included such topics as cooking and embroidery, and at age 13 she won prizes at a fair for both an embroidered handkerchief and the best loaf of bread [2, 4].

Her family moved from Dunstable, Massachusetts, to Westford, Massachusetts, where she attended the Westford Academy, which was regarded as a fine school. This was at a time when public education was not available or mandatory and many women did not receive what we regard today as a fundamental education. Her father opened a store to earn more money that could be used for her education, and Richards attended to her studies and helped in the store. She also tutored other students and began collecting plants and fossils. Her aptitude for applied science became apparent, and she was able to observe the products that women bought at her father's store leading to an awareness of product purity as well as water and air contamination [2, 4].

Richards graduated from the Westford Academy in 1863 and began working diligently to save money at whatever opportunities came her way including teaching, nursing, housekeeping, cleaning, and working in a store to further her education, all at a time when her mother's ill-health took a tremendous toll on her. During the particularly hard last 2 years before she went to Vassar, which she labeled as "Purgatory," she wrote a cousin expressing her hope about the future: "Pray for me, dear Annie, that my life may not be entirely in vain, that I may be of some use in this sinful world." After Vassar College in Poughkeepsie, New York, opened in 1865, she was admitted as a student and was able to attend starting in September 1868 (at age 25) with the 300 dollars that she had saved [1–5].

Richards was admitted as a special student, joined the senior class in her second year, and earned a bachelor's degree in 1870. At Vassar, she seriously considered a major in astronomy as she was significantly influenced by the groundbreaking scientist and advocate for women's education Prof. Maria Mitchell. Prof. Mitchell, who taught astronomy as the first professor hired at Vassar, was the first woman elected to the American Academy of Arts and Sciences. However, Richards majored in chemistry which she thought would be a subject that would have more practical applications than astronomy. She also tutored some of the younger students in mathematics and Latin, helping her to earn the money she needed while at Vassar. On a personal note, she was known around campus for not wasting a moment—she knitted while walking up the five flights of steps to her dorm room and read books while she walked to class! When she graduated, still not sure about the direction of her life, she wrote to her parents: "My life is to be one of active fighting" [2, 4, 5].

In the fall of 1870, Richards was accepted to MIT, which had opened in 1861, as a "special student" labeled the "Swallow Experiment." As the first woman admitted to a scientific school in the USA, Richards was not expected to pay tuition (she thought because of her financial situation), "but I learned later it was because he [the President] could say I was not a student, should any of the trustees or students make a fuss about my presence. Had I realized upon what basis I was taken, I would not have gone."[1] She was careful not to "roil" the waters, but she fervently wished to keep education options open for women who would follow her: "I hope in a quiet way I am winning a way which others will keep open" [4].

In 1872, she began her work on water testing through her association with Professor William Nichols, whose consulting work with the Massachusetts Board of Health included testing public water supplies. Nichols, who had objected to admitting women to MIT, said that the studies Richards conducted "made her a preeminent international water scientist even before her graduation" (and she was his best student!). She also worked with industrial chemist, Professor John Ordway. In 1873, she was awarded a B.S. degree from MIT (see Appendix A for a transcription of her thesis) as well as an M.A. degree from Vassar College. The master's degree from Vassar was based on her estimation of vanadium in iron ore samples [2, 3] (Fig. 1.1).

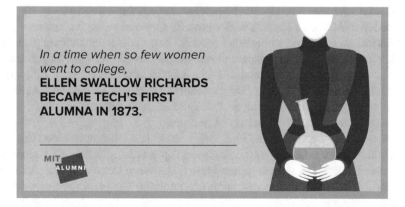

In a time when so few women went to college,
ELLEN SWALLOW RICHARDS BECAME TECH'S FIRST ALUMNA IN 1873.

MIT
ALUMNI

Fig. 1.1 Ellen Swallow Richards Acknowledgement (Source: MIT Alumni Association)

---

[1] Women were allowed to enroll officially at MIT in 1876.

She remained at MIT for 2 more years as a graduate student but did not receive a doctorate as the school did not want the first person awarded a doctorate in chemistry to be a woman. During those years, her interest in practical chemistry continued to increase [1, 2].

Ellen Swallow married Professor Robert Hallowell Richards in 1875. He was a professor of mining engineering and head of MIT's metallurgical laboratory, where she worked for 10 years to further the cause of scientific education for women. During this time, she consulted for her husband by "keeping up with all the German and French mining and metallurgical periodicals (some 20 papers a week come into our house)" and in 1879 became the first woman elected into the American Institute of Mining and Metallurgical Engineers. Her analysis of copper ore samples from the Copper Cliff Mine in Ontario, Canada, led to the development of a significant nickel industry there [1, 2, 4, 6].

In 1876, the Women's Education Association of Boston provided the funding she had requested to open the Woman's Laboratory at MIT. She was already familiar with this organization as they had provided funding for a course she had helped teach as an undergraduate: an experimental course in chemistry at the Girls' High School in Boston. Her new Laboratory, in space provided by MIT and with apparatus, books, and scholarships provided by the Women's Education Association, provided women with training in chemical analysis, industrial chemistry, biology, and mineralogy. Professor Ordway was in charge of the Laboratory; Richards was officially his assistant. She contributed her own funds annually during the years the Laboratory was open. She focused individual attention on the women students ensuring that they had all necessary remedial instruction and advised them on their health and finances. Many of the approximately 500 women who were trained there were able to secure high-level industrial and government consulting jobs. The Laboratory was ultimately so successful that it was closed in 1883, as four women had graduated and the rest were admitted to the regular courses at MIT [1–5].

Richards also performed consulting assignments for both government and industry based on the work and clients she had met through Professor Ordway while she directed the Woman's Laboratory. Her projects included testing commercial products including home furnishings and food for toxic contaminants (wallpaper and fabrics for arsenic) and air, water, and soil for the presence of harmful substances. She also worked with commercial oils for which she developed tests for impurities (referred to as the world standard for the evaporation of volatile oils) and studied the causes of spontaneous combustion. Through her work with noncombustible oils, she became an authority on industrial and urban fires. She worked with Edward Atkinson to design the Aladdin Oven, which led to fire-resistant factories that were models for industry. Her colleague Atkinson appointed her as the first woman science consultant to the Manufacturers Mutual Insurance Company. She was also a sought-after expert on analyzing schools and public buildings for fire resistance [1].

Her students at the Woman's Laboratory assisted in the analyses of impurities of such common household staples as soda, vinegars, and washing powders. They found watered-down milk, samples that were supposed to be cinnamon that were entirely mahogany sawdust, sugar that was contaminated with salt and sand, and sauces with tainted meat. These analyses became the foundation for her books

The Chemistry of Cooking and Cleaning (1882) and *Food Materials and Their Adulterations* (1885). The first Massachusetts Pure Food and Drug Act was enacted in 1882 due to her pioneering research on public health. Richards consulted for the US Department of Agriculture for whom she lectured and authored books, papers, and bulletins on nutrition [1, 2, 4, 5, 7, 8].

Richards also organized the science section of the Society to Encourage Studies at Home at this time. Since women were still not being welcomed into many colleges or universities, this correspondence school provided a means for women to obtain a scientific education. Through the science section, she influenced thousands of women who were being exposed to science for perhaps the first time and sent them microscopes, texts, lesson plans, and specimens. She encouraged them to study anything that interested them whether it be plants, food, or water. One student wrote "I have eyes to see what I never saw before." She personally communicated with many of the students and began to realize the universality of ill-health among middle-class housewives. To this end, she began stressing the importance to her students of healthful food, comfortable dress, and exercise—both physical and mental—and prepared a pamphlet to this end [2, 4, 5].

Richards again prevailed upon the Women's Education Foundation in 1881 to fund the Summer Seaside Laboratory for the purpose of conducting research in the field of marine biology, an undeveloped academic field at that time. Originally in Annisquam, Massachusetts, in 1887, the Summer Seaside Laboratory moved to Woods Hole, Massachusetts, and is now known as the famous Marine Biology Laboratory at Woods Hole. An amazing part of her legacy is that the remaining pioneering women in water quality profiled in this chapter—Ruth Patrick, Rachel Carson, and Sylvia Earle—all had affiliations with the Woods Hole Oceanographic Institution (founded in 1930) [5].

In 1882, she was instrumental in organizing the Association of Collegiate Alumnae (today known as the American Association of University Women—AAUW). The organization was founded to fight the prevailing opinion that too much education was dangerous for women as it was deemed to be dangerous to their health. The organization provided fellowships to women in order to ensure that they had the means to fund their education and worked to raise educational standards at the college level for women. Richards was a leader in efforts to improve physical education in colleges and to widen educational opportunities for women, particularly for graduate education [2–5].

Richards was appointed as an instructor of sanitary chemistry, after the Woman's Laboratory closed—a position she would hold for the rest of her life. This position was located in MIT's newly established laboratory for the study of sanitation chemistry and engineering with Professor Williams Nichols as the director and Ellen Swallow Richards as his assistant. She taught air, water, and sewage analysis and treatment and introduced biology to MIT's curriculum after MIT established the first program in sanitary engineering anywhere in the USA. Her textbook *Air, Water and Food for Colleges* was published in 1900 with A.G. Woodman. Her male students became the designers and operators of the world's first modern sanitation facilities, and she encouraged them to become "missionaries for a better world" [1, 2, 4].

She supervised a study in 1887–1889 of the quality of Massachusetts' inland waters for the Massachusetts State Board of Health. Over 40,000 samples of water were analyzed which represented the water supply consumed by 83% of the state's population. In total, 100,000 analyses were conducted—a study that was unprecedented in its scale. This effort and her involvement with environmental chemistry were significant contributions to the new science of ecology, for which she is often referred to as the "mother." She helped invent new laboratory techniques and apparatuses which were needed to conduct these new analyses. Richards actually plotted the samples on a map of the state of Massachusetts. Through these efforts, she was able to detect geographic patterns in the chlorine data. Thus, she developed the "Normal Chlorine Map" showing levels of similar chlorine throughout the state via isochlors (imaginary lines linking the places that have the same levels of chlorine) that served as an early warning system for inland water pollution.[2] She also developed "Water Purity Tables," that were the first state water quality standards developed in the USA. She was the official water analyst for the Massachusetts State Board of Health for the next 10 years [1–4, 9] (Figs. 1.2 and 1.3).

Richards opened the New England Kitchen in Boston in 1890 to demonstrate how to select and prepare food in a wholesome manner. This effort was the result of a study that she conducted, funded by Pauline Agassiz Shaw, "of the food and nutrition of working men and its possible relation to the question of the use of intoxicating liquors." This kitchen offered cooked food scientifically prepared for maximum nourishment for sale to be eaten at home. The cooking area was also open for public viewing to demonstrate how that food had been prepared. In 1893, Richards set up the Rumford Kitchen at the Chicago World's Fair that was an extension of the New England Kitchen. She expanded the reach of her food preparation knowledge to the

Fig. 1.2 Ellen Swallow Richards' Normal Chlorine Map

---

[2] Significant deviations from normal levels of chlorine are an indication of sewage contamination.

Fig. 1.3 Ellen Swallow Richards collecting water samples. (Courtesy MIT Museum)

Boston School Committee when it contracted with the New England Kitchen to provide nutritious school lunches. Her work led to other school systems and hospitals consulting with her and seeking her expert advice on diet and nutrition [4].

She organized the first of a series of conferences in 1899 at Lake Placid, New York, "for the betterment of the home." These conferences are credited with formally establishing the profession of home economics. The conferences created courses of study for the new field at all levels of instruction including public schools, colleges, agricultural and extension schools, and women's clubs; prescribed standards for teacher training; compiled bibliographies of references; and discussed nutrition, sanitation, and hygiene [4].

She also helped organize a school of housekeeping in 1899 at the Woman's Educational and Industrial Union in Boston that became the department of home economics at Simmons College. Richards was instrumental in founding the American Home Economics Association established for the "improvement of living conditions in the home, the institutional household and the community." She insisted that this new field of home economics be based on economics and sociology and served as the first president until her retirement in 1910. She even helped financially sponsor the Association's *Journal of Home Economics* and wrote 10 books applying science to issues of daily life. Her home served as a living laboratory: there were no heavy draperies, carpets, or coal stove. Instead, she had lightweight curtains, rugs over hardwood floors, a vacuum cleaner, gas for cooking, ventilation, year-round hot water heater, and a telephone. She applied the principles of a clean environment to homes and buildings, addressing what today is often referred to as "Sick Building Syndrome." She undertook a survey of the buildings at MIT and informed the

President that many had serious health hazards that needed to be remedied and for which she recommended solutions. She believed that "One of the most serious problems of civilization is maintaining clean water and clean air, not only for ourselves but for the Planet" [1].

Richards is also credited with discovering the process of cleaning wool with naptha through which she revolutionized the dry cleaning industry. She examined methods to systematize and simplify housework and feed families at reasonable costs. In 1910, the council of the National Education Association appointed her to supervise the teaching of home economics in public schools. Truly, for 25 years, Richards had been "the prophet, interpreter, inspirer, and the engineer" of home economics [2–4, 8].

Richards was a prolific author and speaker and a person who believed that knowledge should be shared. She authored more than 30 books and pamphlets and published numerous papers in addition to the nutrition bulletins she produced for the US Department of Agriculture. Some of her books, in addition to those listed above, include *Home Sanitation: A Manual for Housekeepers* (1887), *Domestic Economy as a Factor in Public Education* (1889), *Sanitation in Daily Life* (1907), *Laboratory Notes on Industrial Water Analysis: A Survey Course for Engineers* (1908), *Conservation by Sanitation* (1911), and *Euthenics: The Science of Controllable Environment* (1912). Richards believed that her ideas could be called euthenics—the science of controlled environment for right living; thus many call her the developer of sanitary engineering. A generous woman with many interests, her sister-in-law referred to her as "Ellencyclopedia" [1, 3, 8].

Ellen Swallow Richards never received a Ph.D. from MIT, but she did receive an honorary Doctor of Science from Smith College in 1910. She was listed in the first edition of *American Men and Women of Science* and was elevated to the rank of Fellow of the American Association for the Advancement of Science (AAAS) in 1878. Richards was a member of the American Chemical Society, the American Public Health Association, and the Boston Society of Natural History. She served for many years on the Board of Trustees of Vassar College. Richards was posthumously inducted into the National Women's Hall of Fame in 1993. She was driven to serve society, and once rued that there were only 24 hours in a day: "I wish I were triplets." The Ellen Swallow Richards Professorship Fund, established in 1973 at MIT on the 100-year anniversary of her graduation, honors her achievements and is intended to strengthen the role of women on the faculty at MIT [5, 7].

1.3 Ruth Patrick (1907–2013)

Dr. Ruth Patrick is credited with laying the groundwork for modern water pollution control efforts. Over her 60-year career, Ruth Patrick advanced the field of limnology, which is the study of freshwater biology. Patrick is recognized, along with Rachel Carson, as having ushered in the current concern for the environment and ecology [7, 10] .

Patrick was encouraged by her father, who gave her a microscope when she was 7 years old and told her "Don't cook, don't sew; you can hire people to do that. Read and improve your mind." [3]. In 2004, she said [10]:

> I collected everything: worms, mushrooms, plants, rocks. I remember the feeling I got when my father would roll back the top of his big desk in the library and roll out the microscope. He would make slides with drops of the water samples we had collected, and I would climb up on his knee and peer in. It was miraculous, looking through a window at a whole other world.

She studied botany, receiving her undergraduate degree from Coker College and both her M.S. (thesis title: "A Study of the Diatoms of Charlottesville and Vicinity") and Ph.D. (dissertation title: "A Taxonomic and Distributional Study of Some Diatoms from Siam and the Federated Malay States") degrees at the University of Virginia. Her summers were spent at the Woods Hole Biological Laboratory, the Cold Spring Harbor Biological Laboratory, and the Biological Laboratory of the University of Virginia at Mountain Lake [11].

Patrick was originally hired as a "volunteer" (without pay—as women scientists at the time were not paid) in 1933 at the Academy of Natural Sciences in Philadelphia, Pennsylvania. Patrick's initial efforts were in microscopy to work with their collection of diatoms, considered to be one of the best collections in the world. Diatoms are microscopic, symmetric single-celled algae with silica cell walls. They are an important part of the food chain of freshwater ecosystems and indicators of water quality. She continued without pay until 1945 while supporting herself through part-time teaching at the Pennsylvania School of Horticulture and making chick embryo slides for Temple University [3, 10, 12] (Figs. 1.4 and 1.5).

She progressed through several positions at the Academy of Natural Sciences including curator of the Leidy Microscopical Society (1937–1947) and the associate curator of the Academy's microscopy department (1937–1947). In 1947, she became the curator and chairwoman of the Academy's limnology department, which she founded, today called the Patrick Center for Environmental Research.[3]

Fig. 1.4 Example centric diatom: Arachnoidiscus sp. (source: IDW-online.de)

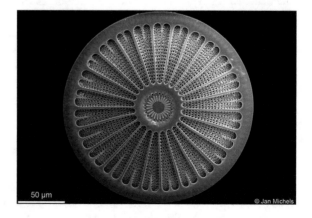

50 µm

© Jan Michels

---

[3] In 2011, the Academy of Natural Sciences became affiliated with Drexel University [13].

Fig. 1.5 Example pennate
diatom: Fragilaria sp.
(source: unh.edu)

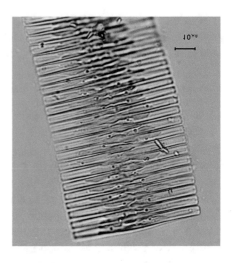

In 1973, Patrick was named the Francis Boyer Research Chair of Limnology at the Academy. From 1973 to 1976, she served as chairwoman of the Academy's board, the first woman to hold that position. Concurrently, she taught at the University of Pennsylvania. Her courses included limnology, pollution biology, and phycology. Her research included taxonomy, ecology, the physiology of diatoms, the biodynamic cycle of rivers, and the diversity of aquatic ecosystems [3, 7, 10–12].

Patrick gave a paper in the late 1940s at a scientific conference on her diatom research in the Poconos. An oil company executive in the audience was so impressed with the ability of diatoms to predict the health of a body of water that he provided the funds to support her research. With these funds, in 1948, Patrick undertook a survey of the then severely polluted Conestoga Creek in Pennsylvania: the first study of its kind. The Creek contained many types of pollution including fertilizer runoff, sewage, and waste products (some of them toxic) from industries in the area. She matched the types and numbers of diatoms in the water to the type and extent of pollution. This procedure is today used universally but was groundbreaking at the time. To aid in the effort, she invented the diatometer, a clear acrylic device that holds glass microscope slides. The diatometer collects the diatoms from bodies of water: they attach to the slides and grow there. Her research showed that healthier bodies of water contain many species of organisms. The belief that biodiversity (the number and kinds of species) is the key indicator of water health is today known as the Patrick Principle in her honor. The Patrick Principle is the foundation of all current environmental assessments [3, 10, 11, 14].

> My great aim has been to be able to diagnose the presence of pollution and develop means of cleaning things up. [15]

Patrick was actively involved in the drafting of the federal Clean Water Act, passed in 1972. She was called the foremost authority on America's river systems. Patrick estimated at one point that she had waded into 850 different rivers around the globe including the Amazon River. She established baseline conditions on the Savannah

River before the opening of the Savannah River Nuclear Power Plant and then monitored river conditions after it opened. At the request of the Army Corps of Engineers, she evaluated the effect of channelization on rivers throughout the USA. After the Three Mile Island accident in 1979, she was asked to assess radionuclide contamination of the Susquehanna River [11].

She was instrumental in 1967 in the founding of the Stroud Water Research Center on the banks of White Clay Creek in Pennsylvania. The first project of the Center was devoted to understanding the biological structure and ecological functioning of a riffle-pool sequence in this creek. The interdisciplinary team approach examined chemistry, microbial ecology, entomology, hydrology, and fish, among other topics, and produced a new paradigm for flowing water systems called "The River Continuum"—one of the most highly cited papers in the field of stream ecology. The Center continues in operation today adhering to Patrick's original vision: to advance knowledge and stewardship of freshwater systems through global research, education, and restoration [11].

Patrick taught at the University of Pennsylvania for more than 35 years and wrote books and hundreds of technical articles. She collaborated with Charles Reimer on the two-volume *Diatoms of the United States*. She also wrote *Groundwater Contamination in the United States* (1983, 1987), *Power So Great, Colors of Tomorrow*, and the series *Rivers of the United States* [3, 7, 10, 12, 16].

Dr. Patrick was the first woman to serve on the board of directors of the DuPont Corporation and was its first environmental activist. She also served on the board of directors of Pennsylvania Power and Light and advised presidents Lyndon B. Johnson (on water pollution) and Ronald Reagan (on acid rain) as well as several Pennsylvania governors on water quality issues. She served on water pollution and water quality panels for the National Academy of Sciences and the US Department of Interior as well as other federal advisory groups. These included the National Academy of Sciences Committee on Science and Public Policy, the General Advisory Committee of the Environmental Protection Agency, the Advisory Council of the Renewable Resources Foundation, and the Smithsonian Institution Council. Her professional affiliations included the Phycological Society of America (president), American Society of Naturalists (president), Botanical Society of America, American Society of Limnology and Oceanography, American Institute of Biological Sciences, Ecological Society of America, and the American Society of Plant Taxonomists. She also served on the board of directors of the World Wildlife Fund [11, 16].

Patrick was a great believer in scientists, government, and industry working together to solve problems. She said: [3, 10, 12]

My great theme in life is that academia, government, and industry have got to work closely on all the big problems of the world. Unless academics and industry get together there will not be many bright young people trained in the future. We have to develop an atmosphere where the industrialist trusts the scientist and the scientist trusts the industrialist. You've got to trust people.

Patrick was elected to the National Academy of Sciences in 1970 as the 12th woman to receive this form of recognition. Patrick received the National Medal of Science in 1996 from President Bill Clinton "for her algal research, particularly the ecology and paleoecology of diatoms, and for elucidating the importance of biodiversity of aquatic life in ascertaining the natural condition of rivers and the effects of pollution." She was elected as a Fellow of the American Academy of Arts and Science in 1976 and was the recipient of over 25 honorary degrees. Dr. Patrick was inducted into the National Women's Hall of Fame in 2009 [2, 10, 14].

1.4   Rachel Carson (1907–1964)

Biologist Rachel Carson is most famous for her 1962 book *Silent Spring*, which exposed the dangers of the overuse of pesticides. Her book is often credited with being the catalyst that led to the first Earth Day in 1970 and the current environmental movement. She hadwrote three other books related to the oceans, of which two became bestsellers.

Carson was determined to be a writer from the time she was a child. She majored in English at the Pennsylvania College for Women where she graduated magna cum laude. While in college, she also found an interest in natural history that was magnified by a required zoology course. She attended graduate school at Johns Hopkins University after spending the summer at the Marine Biological Laboratory in Woods Hole, Massachusetts. In 1932, she earned her master's degree in marine biology from Johns Hopkins University [2, 7].

Carson decided to apply for and accepted (in 1936) a position as a junior aquatic biologist with the US Bureau of Fisheries, for whom she had previously worked on a part-time basis, in order to help support family members. She was one of the first two women to be hired by the Bureau in a professional position. She would stay a federal government employee through 1952 [2, 7].

Her first article "Undersea" was published in 1937 in the *Atlantic Monthly*. Her first book *Under the Sea Wind—a Naturalist's Picture of Ocean Life* was published in 1941. Although well- received, it did not become a bestseller until after the 1951 publication of *The Sea Around Us*. *The Sea Around Us* was not only a bestseller; it won a National Book Award and was translated into 30 languages [6, 7, 17].

The Bureau of Fisheries merged with the Biological Survey in 1940 to form the US Fish and Wildlife Service, with a stated purpose of conservation. Carson wrote two large monographs during World War II, *Food from the Sea: Fish and Shellfish of New England* (1943) and *Food from the Sea: Fish and Shellfish of the South Atlantic and Gulf Coasts* (1944). The goals of these monographs were to prevent overfishing of certain fish species and to help find new food sources. In 1947, Carson became editor in chief of the bureau's publications. She produced 12 pamphlets regarding national wildlife refuges in the series titled *Conservation in Action* [6, 7].

Carson enjoyed some success with the 1951 publication of *The Sea Around Us* and her receipt of a Guggenheim Foundation fellowship. She felt financially secure enough to resign from her job and turned to writing full time. She told a friend "If I could choose what seems to me the ideal existence, it would be just to live by writing." She had previously said that "biology has given me something to write about." Her next book, *The Edge of the Sea*, was published in 1956. These books

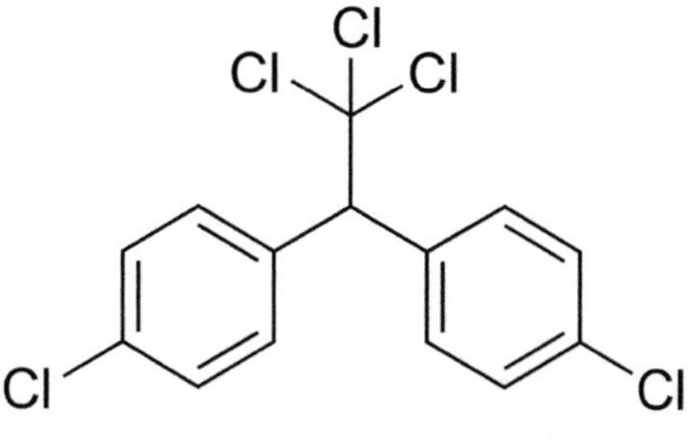

Fig. 1.6 DDT
(1,1,1-trichloro-2,2′-di(4-
chloropnehyl)ethane

gave lyrical and precise information on the physics, chemistry, and biology of the ocean and its shores in a way that was both enjoyable to read and educational [2, 7].

Her seminal work, *Silent Spring*, published in 1962, is an exposé on the dangers of the pesticide DDT. A bestseller, the book has been credited with initiating the environmental movement of the 1960s and 1970s and being the catalyst for the first Earth Day. The book raised the concept, not previously considered by the scientific community or the public at large, that human behavior could cause irreparable damage to the environment. It also caused much controversy. Never before had there been the thought that small perturbations in the food chain could cause severe, far-reaching effects. After *Silent Spring*, the world took a new direction [7] (Fig. 1.6).

Rachel Carson was the recipient of multiple awards including the 1952 John Burroughs Medal, the 1954 Gold Medal of the New York Zoological Society, and the 1963 Conservationist of the Year Award from the National Wildlife Federation. She was inducted into the National Women's Hall of Fame in 1973 [7].

1.5 Sylvia Earle (1935–)

Dr. Sylvia Earle has been called "Her Deepness," a "Living Legend," and a "Hero for the Planet." Oceanographer, author, lecturer, and marine biologist Earle was encouraged by her parents in her love of nature. She said:

> I wasn't shown frogs with the attitude 'yuk,' but rather my mother would show my brothers and me how beautiful they are and how fascinating it was to look at their gorgeous eyes. [3]

When it was time for her to go to college, Earle's parents supported her major in biology but wanted her to get her teaching credentials and learn how to type "just in case" [3, 18].

Earle received her B.S. from Florida State University in 1955 and her master's degree in botany from Duke University in 1956. She conducted fieldwork in the Gulf of Mexico as her thesis was a detailed study of algae in the Gulf. Earle said she has collected more than 20,000 samples and still keeps informed on that ecosystem. She said "When I began making collections in the Gulf, it was a very different body of water than it is now—the habitats have changed. So I have a very interesting baseline." She received her Ph.D. from Duke University in 1966 [3].

Earle was the resident director of the Cape Haze Marine Laboratories in Sarasota, Florida, and later moved to Massachusetts and served as both a research scholar at the Radcliffe Institute and research fellow at the Farlow Herbarium at Harvard University. In 1976, she moved to California where she became a research biologist and curator at the California Academy of Sciences and a fellow of botany at the Natural History Museum, University of California, Berkeley [3].

The love of the sea was also calling Earle during these years. In 1970, Earle lived in an underwater chamber for 14 days with four other oceanographers as part of the government-funded Tektite II Project. This was the first all-female team and she was its leader. When they returned to the surface, they had a ticker-tape parade and a reception at the Nixon White House. All told, she has led more than 50 expeditions and logged more than 7000 hours under water. Her dedication to learning about the oceans and their conservation was not always honored by newspaper accounts of her expeditions. One such headline from the Mombasa, Kenya, paper read "Sylvia sails away with 70 men—and expects no problem," and one headline about the Tektite II project read "6 women and only one hairdryer" [3, 19, 20].

The study of marine biology was dramatically changed with the advent of self-contained underwater breathing apparatus (SCUBA) equipment. Earle was one of the first researchers to use SCUBA equipment and observe plant and animal habitats and life beneath the sea and was able to identify many new species. She set the unbelievable record of free diving to a depth of 1250 feet [3].

Earle and her former husband recognized the limitations of SCUBA and set up a company in 1981 to build a submersible craft that could dive deeper than humans in SCUBA gear. The design for the *Deep Rover* submersible was originally sketched on a napkin. The design is still used as a mid-water machine capable of going to ocean depths of up to 3000 feet [3].

Earle was the first woman to serve as the chief scientist for the National Oceanic and Atmospheric Administration (NOAA). NOAA conducts underwater research, manages fisheries, collects atmospheric data, and monitors marine spills. After she left that position, she worked with a team of Japanese scientists to develop equipment to send a remote submersible to depths of 36,000 feet [3] (Fig. 1.7).

Dr. Earle is currently Explorer-in-Residence at the National Geographic Society and the Rosemary and Roger Enrico Chair for Ocean Exploration. Earle says:

I hope for your help to explore and protect the wild ocean in ways that will restore the health and, in so doing, secure hope for humankind. Health to the ocean means health for us. [18]

Fig. 1.7 NOAA logo

Fig. 1.8 LEGO figure of
Dr. Sylvia Earle. (Maia
Weinstock used with
permission)

It's filled with living things that shape the chemistry of Earth. It's where we get most of our oxygen, from small organisms that make up the plankton in the sea. Of course trees and other green things have much to do with generating oxygen and taking up carbon dioxide— but there's no green without blue. [21]

Earle has published books and hundreds of scientific papers and other publications on marine life. She has been at the helm of ocean exploration for more than four decades. Earle is a dedicated advocate of public education regarding the importance of oceans as an essential environmental habitat. To promote ocean education, LEGO issued a Dr. Sylvia Earle LEGO set in 2015 including a minifig of Dr. Earle as well as a deep-sea exploration vessel, deep-sea submarine, deep-sea helicopter, deep-sea starter set, and a deep-sea SCUBA scooter [3, 18, 19] (Fig. 1.8).

Earle has received more than 100 national and international honors including Glamour Women of the Year, Netherlands Order of the Golden Ark, Australia's International Banksia Award, and medals from the Royal Geographical Society, the National Wildlife Federation, and the Philadelphia Academy of Sciences. Dr. Earle is a tireless advocate for the Earth's oceans and was inducted into the National Women's Hall of Fame in 2000 [3, 18, 19].

All the authors in this volume are women who have contributed to water quality science and research. Enjoy learning about their work and reading their stories!

References

1. Weingardt RG (2004) Engineering legends: Ellen Henrietta Swallow Richards and Benjamin Wright. Leadersh Manage Eng: ASCE 4:156–160
2. Ogilvie M, Harvey J (eds) (2000) The biographical dictionary of women in science: pioneering lives from ancient times to the mid-20th century. Routledge, New York
3. Profitt P (ed) (1999) Notable women scientists. The Gale Group, Detroit
4. James ET (1971) Notable American women: a biographical dictionary. The Belknap Press of Harvard University Press, Cambridge
5. Durant E MIT technology review, "Ellencyclopedia." https://www.technologyreview.com/s/408456/ellencyclopedia/. August 15, 2007

6. Kass-Simon G, Farnes P (eds) (1990) Women of science: righting the record. Indiana University Press, Bloomington

7. Bailey MJ (1994) American women in science a biographical dictionary. ABC-CLIO, Denver

8. Shearer BF, Shearer BS (eds) (1997) Notable women in the physical sciences: a biographical dictionary. Greenwood Press, Westport

9. https://thisdayinwaterhistory.wordpress.com/tag/ellen-swallow-richards/

10. http://www.nytimes.com/2013/09/24/us/ruth-patrick-a-pioneer-in-pollution-control-dies-at-105.html

11. Bott TL, Sweeney BW (2013) A biographical memoir, Ruth Patrick 1907–2013, National Academy of Sciences. http://www.nasonline.org/publications/biographical-memoirs/memoir-pdfs/patrick-ruth.pdf

12. Shearer BF, Shearer BS (eds) (1996) Notable women in the life sciences: a biographical dictionary. Greenwood Press, Westport

13. http://www.ansp.org/about/drexel-affiliation/

14. https://www.nationalmedals.org/laureates/ruth-patrick

15. Zauzmer J "Ruth Patrick, ecology pioneer, dies at 105." https://www.washingtonpost.com/national/health-science/ruth-patrick-ecology-pioneer-dies-at-105/2013/09/23/2bcde762-245e-11e3-b75d-5b7f66349852_story.html?utm_term=.5d9dd8392428. September 23, 2013

16. The Academy of Natural Sciences of Drexel University, A Biography of Ruth Patrick. http://www.ansp.org/research/environmental-research/people/patrick/biography/

17. Turkson N The Atlantic in nature. http://www.theatlantic.com/notes/2016/02/atlantic-nature-writing/460158/. February 9, 2016

18. National Geographic, Explorers: Bio – Sylvia Earle. http://www.nationalgeographic.com/explorers/bios/sylvia-earle/

19. TED: Ideas worth spreading, Sylvia Earle – Oceanographer. https://www.ted.com/speakers/sylvia_earle

20. http://protecttheoceans.org/wordpress/?p=1459

21. http://variety.com/2014/film/features/dr-sylvia-earle-who-leonardo-dicaprio-calls-an-inspiration-1201283987/

Jill S. Tietjen, P.E., entered the University of Virginia in the fall of 1972 (the third year that women were admitted as undergraduates) intending to be a mathematics major. But midway through her first semester, she found engineering and made all of the arrangements necessary to transfer. In 1976, she graduated with a B.S. in Applied Mathematics (minor in Electrical Engineering) (Tau Beta Pi, Virginia Alpha) and went to work in the electric utility industry. During her 40 years in the industry, she worked for Duke Power Company, Mobil Oil Corporation's Mining and Coal Division, Stone & Webster Management Consultants, RCG/Hagler Bailly, and for her own firm, Technically Speaking, Inc. She also served as the Director of the Women in Engineering Program at the University of Colorado at Boulder from 1997 to 2000.

Galvanized by the fact that no one, not even her Ph.D. engineer father, had encouraged her to pursue an engineering education and that only after her graduation did she discover that her degree was not ABET-accredited, she joined the Society of Women Engineers (SWE) and for 40 years has worked to encourage young women to pursue science, technology, engineering, and mathematics (STEM) careers. In 1982, she became licensed as a professional engineer in Colorado. She served as National President of SWE during 1991–1992.

Tietjen started working on jigsaw puzzles at age 2 and has always loved to solve problems. She derives tremendous satisfaction seeing the result of her work—the electricity product that is so reliable that most Americans just take its provision for granted. Flying at night and seeing the lights below, she knows that she had a hand in this infrastructure miracle. An expert witness, she worked to plan new power plants.

Her efforts to nominate women for awards began in SWE and have progressed to her acknowledgment as one of the top nominators of women in the country. Her nominees have received the National Medal of Technology and the Kate Gleason Medal; they have been inducted into the National Women's Hall and Fame and state halls including Colorado, Maryland, and Delaware and have received university and professional society recognition. Tietjen believes that it is imperative to nominate women for awards—for the role modeling and knowledge of women's accomplishments that it provides for the youth of our country.

Tietjen received her MBA from the University of North Carolina at Charlotte. She has been the recipient of many awards including the Distinguished Service Award from SWE (of which she has been named a Fellow) and the Distinguished Alumna Award from the University of Virginia, and she has been inducted into the Colorado Women's Hall of Fame. Tietjen sits on the boards of Georgia Transmission Corporation and Merrick & Company. Her publications include the bestselling and award-winning book *Her Story: A Timeline of the Women Who Changed America* for which she received the Daughters of the American Revolution History Award Medal. Her latest book, published in 2019, is *Hollywood: Her Story, An Illustrated History of Women and the Movies*.

Part II
Water Quality in Engineered Systems

Chapter 2
Integrating Engineered and Nature-Based Solutions for Urban Stormwater Management

Laura A. Wendling and Erika E. Holt

Abstract Urban areas increasingly face the challenge of effectively managing water resources to minimize both flooding and freshwater scarcity. Hydrometeorological consequences of climate change exacerbate the effects of surface sealing and increased runoff in urban areas, the overexploitation of available water resources, water pollution, and aging infrastructures. These issues highlight the need for new robust and reliable techniques to manage flooding and improve the quality of surface runoff. The effective integration of robust engineering and design standards, novel material technologies, and innovative blue-green infrastructure solutions can serve to reconnect the urban hydrologic cycle, enhancing the resilience of urban areas to climate change. Engineered blue-green-gray systems that combine urban waterways with functional vegetation, geo- or bio-based filter materials, and related technologies can create holistic systems for sustainable management of urban stormwater quantity and quality.

2.1 Introduction

Urban stormwater runoff threatens the ecological integrity of urban waterbodies but also provides an underutilized opportunity to capture, treat, and use stormwater in urban applications. Stormwater runoff across surfaces and sewer network overflows are increasingly common in cities due to the increasing extent of impervious surfaces, aging infrastructure, and often undersized, centralized stormwater networks coupled with global changes in precipitation intensity. Sealing of soil surfaces by impervious materials is a common consequence of urbanization. Constructed surfaces such as rooftops, sidewalks, roads, and parking lots covered by impenetrable materials prevent water from infiltrating the underlying soil. Sparse vegetation on impervious surfaces increases the speed at which water is able to move across these surfaces relative to a vegetated landscape. Thus, surface sealing substantially alters

L. A. Wendling (✉) · E. E. Holt
VTT Technical Research Centre of Finland, Espoo, Finland
e-mail: laura.wendling@vtt.fi; erika.holt@vtt.fi

© Springer Nature Switzerland AG 2020   23
D. J. O'Bannon (ed.), *Women in Water Quality*, Women in Engineering
and Science, https://doi.org/10.1007/978-3-030-17819-2_2

Fig. 2.1 Surface sealing in
urban areas reduces
infiltration (near field),
increasing stormwater
runoff and flood risk

the urban hydrologic cycle by reducing infiltration and increasing the surface
stormwater flowrates (Fig. 2.1). The net result of increased surface imperviousness,
or surface sealing, in urban areas is increased stormwater runoff volume and rates of
peak stormwater discharge, along with increased pollutant loads to receiving
waterbodies.

There are many hydrometeorological consequences of climate change. Changes
in rainfall exacerbate the effects of surface sealing and increased runoff in urban
areas, highlighting the need for new robust and reliable ways to manage flooding
and the quality of surface runoff. Flood events are expected to increase in the future
as a consequence of climate change and are particularly costly, accounting for two-
thirds of the economic costs of damages attributed to natural disasters in Europe
[11]. Nearly 20% of European cities with >100, 000 inhabitants are highly vulnerable to flooding [12].

Urban stormwater collection systems are burdened by increasing urban populations in addition to the increased proportion of stormwater runoff compared with
undeveloped landscapes. The world's population has gradually transitioned from
predominantly rural to urban, with the urban population exceeding the rural population worldwide for the first time in 2007. Urbanization is particularly marked in
postindustrialized nations of the Northern hemisphere with approximately 73% and
81% of the European and North American populations, respectively, residing in
urban areas in 2014 [35]. The United Nations [35] predicts that, by the year 2050,
66% of the world's population will be living in urban areas.

Urban stormwater contains a variety of contaminants that can adversely affect
receiving waters. High sediment loads and contaminants such as metals, pathogens,
and pesticides as well as the nutrient elements nitrogen and phosphorus are common in surface runoff. Most stormwater runoff is discharged directly into rivers,
lakes, and the sea without any treatment. Only that portion of stormwater captured
by the urban sewer network can be treated in municipal water treatment facilities.

The intermittent presence of stormwater in combined sewer networks, typical in older town centers and some industrial estates, can result in storm event wastewater overflows. Municipalities face the complex challenge of effectively managing water resources to minimize both flooding and the discharge of pollutants associated with the higher volumes of stormwater runoff to improve environmental sustainability and safeguard the well-being of urban residents.

The composition and quantities of contaminants in urban stormwater are largely dictated by catchment land use. Road surfaces are the principal source and conveyor of pollutants in urban stormwater runoff [24], but urban stormwater runoff also includes surface runoff and drainage discharges from other impervious surfaces as well as public open spaces and green spaces, road verges, and construction sites (Table 2.1). Metals and substances that contribute to eutrophication of surface waters (in particular, nitrogen and phosphorus) are among the main pollutants of concern listed in the EU Water Framework Directive [7]. Metals are particularly prevalent in urban dust and surface runoff. Vehicle traffic is a major source of cadmium, copper, lead, nickel, and zinc [1, 15, 16, 39]. The EU Water Framework

Table 2.1 Pollutant sources, mean event concentrations, and European Union and US environmental water quality standards. All pollutant concentrations are expressed in mg/L. (Modified from Lundy et al. [24])

Pollutants and sources	Event mean concentrations	EU water quality standards[a, b]	US water quality standards[c-e]
Sewer misconnections; urban amenity fertilizer; residential; highways, motorways and major roads; road gullies and pipe drains	Total P: 0.02–39	Total P: 0.18	Total P: 0.008–0.128[c]
	Total N: 0.1–20	Total N: 5–65	Total N: 0.1–2.18[c]
	NO_3: 0.1–4.7		
	NH_4: 0.2–3.8	NH_4: 0.25–9[f]	
Highways, motorways and major roads; urban distributor roads; suburban roads; commercial estates; residential; roofs; road gullies and pipe drains	Cd: 0.2–13	Cd: AA 0.08–0.25[g]; MAC 0.45–1.5[g]	Cd: CMC 2; CCC 0.25[d, g]
	Pb: 1–2410	Pb: AA 7.2; MAC N/A	Pb: CMC 65; CCC 2.5[d, g]
		Cu: 1–28[g]	Cu: variable, calculated via biotic ligand model
	Ni: 2–493	Ni: AA 20; MAC N/A	Ni: CMC 470; CCC 52[d, g]
	Zn: 53–3550	Zn: 8–125[g]	Zn: CMC & CCC 120[d, g]

[a]European Union environmental quality standards (EQSs) for priority and priority hazardous substances Cd, Pb, Ni [8]; UK EQSs for Cu, Zn [5]
[b]*AA* annual average concentration, *MAC* maximum allowable concentration
[c]Total P and total N: Summary of Ecoregional Nutrient Criteria for freshwater systems [34]
[d]US EPA Recommended Water Quality Criteria for freshwater aquatic life
[e]*CMC* Criteria Maximum Concentration, *CCC* Criteria Continuous Concentration
[f]Depending on designation of use
[g]Freshwater criteria vary depending on water hardness

Directive lists cadmium, lead, and their respective compounds as priority substances, and cadmium is further identified as a priority hazardous substance.

Chemical fertilizers applied to lawns, gardens, public open spaces, golf courses, and other areas are major contributors to nutrient loads in urban stormwater runoff [4, 14, 30, 38]. Excessive nutrient inputs to coastal waters can result in the development of hypoxic zones. Hypoxic zones may occur naturally in coastal marine areas; however, anthropogenic activities during the previous 50 years have resulted in a dramatic increase in the magnitude and extent of hypoxic coastal zones around the world [27]. Marine hypoxic zones develop as a result of nitrogen and phosphorus enrichment of coastal waters. The quantity of reactive nitrogen and phosphorus currently released to the world's oceans is an estimated three times greater than emissions during pre-industrial times [26, 32]. This abundant supply of nutrients causes eutrophication: a tremendous increase in net primary productivity by phytoplankton and other autotrophs. When the autotrophs die, they sink to the bottom of the waterbody and are decomposed by bacteria, resulting in reduced oxygen levels. As winds move surface waters away from the coastline, the hypoxic deeper layers are brought to the surface creating a hypoxic zone. Thus, attenuation of nutrients in aqueous discharges from urban areas is essential to protect or restore coastal aquatic ecosystems.

Traditional stormwater management in urban areas focused predominantly on mitigating flood risk. Combined sewer systems, common in older urban areas, collect both stormwater and wastewater in a single pipe network for conveyance to a central water treatment facility. Surface runoff volumes during large precipitation events can exceed the capacity of combined sewer networks, leading to the discharge of combined stormwater runoff and untreated wastewater. Separate sewer systems comprised of individual pipe networks for stormwater and wastewater, respectively, significantly reduce the potential for wastewater overflows, but localized flooding by stormwater runoff remains a concern. The capacity of a given stormwater sewer network is engineered based on storm frequency analysis using historical precipitation data, stormwater management plans, flood studies, and/or conveyance system analysis. Although engineered systems are highly reliable under the conditions for which they were designed, extreme weather events and increased surface imperviousness in the drainage area can result in surface runoff in excess of the capacity of the stormwater sewer networks.

Increased stormwater runoff in urban areas has traditionally been managed by increasing the number and size of sewer pipes and drainage channels. Sewers, pumps, drainage channels, and sedimentation basins or stormwater retention ponds are classically viewed as gray infrastructure for stormwater management. Replacement or expansion of sewers and drainage channels can be highly effective for runoff attenuation but fail to address the discontinuity of the urban hydrologic cycle presented by surface sealing and limited infiltration. In addition, major subsurface engineering works are costly and may limit or disrupt other underground services. While surface runoff captured by the stormwater sewer network is typically discharged to a receiving surface waterbody without further treatment, the system can be designed to discharge captured stormwater runoff to a wetland or infiltration basin to provide additional water storage and to manage contaminants in stormwater runoff. This type of green-gray stormwater management solution may be employed to expand stormwater sewer network capacity in urban areas.

Pipe networks, reservoirs, treatment plants, and similar human-engineered components are referred to as gray infrastructure. Green infrastructure such as rain gardens, green roofs, permeable pavements, swales, wetlands, and other systems designed to reduce stormwater runoff volume are identified in the EU Soil Sealing Guidelines [10] as stormwater management solutions that enhance urban environments. In combination with blue infrastructure, or urban landscape elements linked to water (lakes, ponds, waterways, etc.), green systems for urban stormwater management have gained popularity due to their cost-effectiveness as well as the multiple co-benefits yielded by blue-green stormwater management systems. Green infrastructure systems often utilize engineered infiltration or subsurface filtration media to optimize hydraulic conductivity/maximize water infiltration, filter particulate pollutants, or provide growth media for microbial communities.

The traditional practice of diverting surface runoff away from urban infrastructure as rapidly as possible can result in increased downstream flood risk, reduced groundwater recharge, and, in the long term, reduced groundwater resource availability as well as potential contamination of receiving waterbodies. A multidisciplinary approach to urban stormwater that effectively integrates robust engineering and design principles with novel material technologies and innovative blue-green solutions can manage stormwater runoff quantity and quality to enhance the resilience of urban areas to climate change.

2.2 Permeable Pavement Systems and Gray-Green Integration

Urban areas experience substantially greater stormwater runoff volume compared with undeveloped landscapes or rural areas. With increasing urban development, rainfall that previously infiltrated the soil is instead increasingly intercepted by impervious surfaces such as rooftops, streets, and parking lots. Water infiltration decreases and more rainfall instead becomes surface runoff as the degree of landscape imperviousness increases (Fig. 2.2).

Permeable pavement systems (PPS) represent a different type of gray infrastructure that facilitates increased infiltration of built surfaces while maintaining the necessary functionality as pathways for vehicles and pedestrians. Green infrastructure elements such as trees, shrubs, bioswales, and rain gardens are frequently combined with PPS to increase cumulative stormwater infiltration, storage, and evaporation, as well as enhance urban amenity value. All PPS have a similar basic structure consisting of a surface pavement layer and an underlying reservoir typically comprised of stone aggregates, which may be underlain by a filter layer or geotextile fabric. Several modifications to this basic design are possible such as the use of different types of subbase materials or the inclusion of water collection zones, pipes, or tanks in connection with impervious layers.

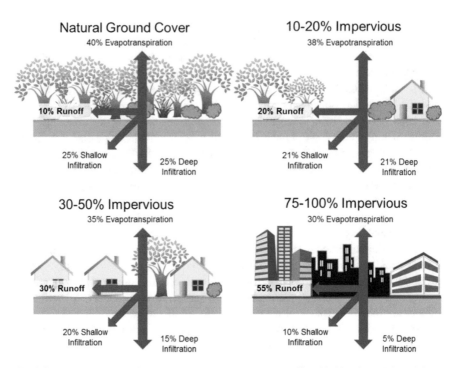

Fig. 2.2 Effects of surface imperviousness on stormwater runoff and infiltration. (Adapted from U.S. EPA [33])

2.2.1 Continuous Permeable Pavements

Permeable, or pervious, concrete (PC) (Fig. 2.3) is a concrete with interconnected pores, typically 1–8 mm in diameter, which account for 20–30% of the total concrete volume [19]. The porosity of PC confers high water and air permeability. Permeable concrete is typically 1600–2000 kg/m^3, approximately 70% the mass of conventional concrete. Relative to conventional concrete, PC shrinks less and has higher thermal insulation value. A typical PC layer is 100–200 mm thick and overlies a granular base or subbase layer. Permeable concrete is considered a stormwater best management practice (BMP) in the USA due to its ability to substantially reduce stormwater runoff. These concretes can be applied similarly to conventional concretes either as a single massive structural unit or as permeable concrete paving blocks, subject to performance limitations.

The use of PC is limited by its compressive, tensile, and flexural strengths, which are lower relative to conventional concrete. In general, there is an inverse relationship between PC porosity and strength. Water permeability typically increases with porosity; however, the pore-size distribution and connectivity of pores are key to permeability such that materials with similar overall porosity and different pore-size distribution or degree of connectivity between pores may exhibit substantially

Fig. 2.3 Water passes easily through the interconnected pores of permeable concrete

different permeability. Freeze-thaw resistance of PC and resistance to clogging as a result of surface grit application can be controlled by managing material pore-size distribution during PC design and construction phases.

Permeable concrete has numerous potential applications in green infrastructure in addition to use as a road or walkway. Drain tiles, greenhouse floors, lightweight noise barriers, floors with enhanced acoustic absorption, artificial reefs, and surfacing for parks and sport courts are additional applications for which PC is well suited. Unreinforced PC is used as the open structure allows air and water ingress; thus caution must be used to minimize risks of corrosion of adjacent metal structures. Mix composition, placement, compaction, and curing can all considerably affect PC function. The rate of water infiltration through a typical PC is 2–5.5 mm/s, and infiltration as high as 20 mm/s has been reported for PC with 30% porosity [19]. Drainage rates between 100 and 750 L/min/m^2 have been reported for PC [31].

Permeable asphalt (PA) has greater porosity relative to conventional hot mix asphalt. Permeable asphalt consists of an open-graded porous mixture with 18–25% porosity and typically 400–500 mm surface thickness [23]. Permeable asphalt exhibits greater skid resistance in wet conditions compared with conventional asphalt due to the absence of a surface water film (Fig. 2.4). Suboptimal material performance, for example, in hot weather and in heavily trafficked areas, may limit the widespread application of PA, but fiber reinforcement can increase PA durability and extend both its application suitability and life span. Aggregate typically comprises 60–90 wt.% of PA, and a high proportion of coarse aggregate is necessary to provide sufficient contact between stones [6]. Binder additives or modifiers are sometimes used to increase PA durability, decrease asphalt viscosity in warm temperatures, and/or increase flexibility in cold temperatures. In general, the greater the strength of the binder, the less is required and thus the greater the porosity of the end product PA.

The thickness of PA depends on traffic loads and can range from 19 to 180 mm. Conventional road structures intended for heavy loads are suitable base materials

Fig. 2.4 A permeable asphalt section of pedestrian and bicycle path in Helsinki, Finland (foreground) shows reduced surface water film formation during rain events compared with conventional asphalt (background)

for PA although care is required to ensure adequate drainage. Like PC, the use of PA may be limited by frost penetration in northern climates, although sufficient drainage of the reservoir base has been shown to minimize issues related to freezing [18, 29].

2.2.2 Interlocking Permeable Pavements

Permeable interlocking concrete pavement (PICP) and permeable natural stone pavement (PNSP) are generally used in pedestrian walkways or residential driveways. These PPS are usually underlain by an open-graded bedding layer of small aggregate. Beneath the bedding course, crushed stones provide a base reservoir ca. 75–100 mm thick. The base reservoir serves as a high infiltration rate layer and a transition zone between the bedding and subbase layers. The subbase generally consists of open-graded 20–65-mm diameter stones, with layer thickness dependent upon water storage requirements and traffic loads. A minimum porosity of about 30% is necessary for water storage in base and subbase layers underlying permeable pavements. An underdrain may be installed within the subbase layer as required for sufficient drainage, particularly where underlying soils have low water permeability.

Water infiltrates PICP and PNSP primarily via joints and/or openings between the pavers that allow water to enter an open-graded aggregate bedding course (Fig. 2.5). The joints between pavers should comprise 5–15% of the total surface area in order to provide sufficient drainage. Joints filled with relatively larger-sized aggregate exhibit a higher infiltration rate relative to those filled with fine-grained aggregate. The selection of joint material is independent on whether the pavers themselves are permeable or impervious blocks. Sand is not recommended as a joint-filling material due to its low infiltration rate and high clogging potential.

2.2.3 Construction and Maintenance Considerations for Permeable Pavements

Permeable pavements are produced using less sand or fines than conventional pavements, yielding stable void spaces that allow water to drain freely. Over-compaction or clogging during construction of PPS can be an issue in the absence of construction oversight and adequate quality controls. Construction surveillance is necessary to ensure base layers of PPS do not become overly compacted by the passage of heavy equipment. Similarly, PPS require protection from high total suspended sediment (TSS) loads during active construction at the site. Land clearing, excavating, grading, and other construction activities can result in significantly increased soil erosion and TSS in site runoff [13, 22, 40]. Most construction guidelines for PPS specify protection from sediment intrusion during active site construction and exclusion of heavy vehicle traffic from permeable pavement areas to prevent clogging and compaction, respectively, during the construction phase.

The choice between different PPS depends on the lifetime duration expectations and loading from both traffic and environmental factors; however, all PPS require some ongoing maintenance for optimum functionality. In time, surface clogging by exogenous fine particulate materials can reduce the porosity and infiltration capacity of PPS if not adequately maintained. In cold climates the use of sand, or gravel containing a high proportion of fines, to provide added traction during winter months can clog the pores of permeable pavements and significantly reduce water infiltration capacity [42]. Vacuuming and power washing of permeable pavements have been shown to restore 39–96% of the original filtration capacity [20],

Fig. 2.5 Water infiltrates permeable interlocking concrete pavement through openings between pavers

demonstrating that routine maintenance to minimize clogging is essential to maintain the infiltration capacity of PPS. Dense vegetation growth in joint filling can reduce infiltration and thus should also be addressed during maintenance activities.

2.2.4 Integration with Green Infrastructure

Permeable pavement systems are frequently utilized in combination with green infrastructure such as trees, rain gardens, green roofs, swales, and wetlands to reduce stormwater runoff volume. Along with blue infrastructure, green-gray systems for urban stormwater management have gained popularity due to their cost-effectiveness as well as the multiple documented co-benefits, such as biodiversity conservation and mitigation of urban heat island effects. Worldwide, the concept of urban drainage has evolved from draining all stormwater as rapidly as possible to a focus on meeting multiple objectives using a series of connected treatment units (Fig. 2.6). This use of multiple connected stormwater treatment units within a single treatment system is sometimes referred to as a treatment train and often includes both pollution source control and treatment practices. Nomenclatures such as low-impact design (LID), sustainable urban drainage systems (SUDS), and water-sensitive urban design (WSUD) systems are all modern, multi-objective stormwater treatment systems.

Design of green-gray stormwater management structures necessitates the purposeful integration of materials science, chemistry, hydrology, plant science, environmental biotechnology, and urban planning knowledge. Although green-gray stormwater management structures have an organic aesthetic of "urban nature," they are, in reality, engineered systems. The long-term well-being of vegetation in urban areas is essential for optimal provision of beneficial services, and the urban setting is a particularly harsh environment for vegetation. Vegetation in green-gray urban runoff treatment structures must survive widely fluctuating levels of inundation,

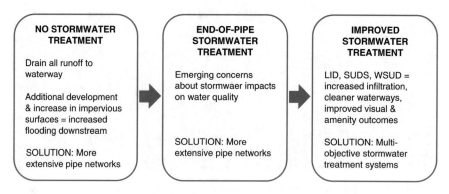

Fig. 2.6 The evolution of modern urban stormwater drainage practices. (Modified from RMS [28])

access to a limited quantity of soil and growth in a confined space, as well as physical stressors, i.e., snow management vehicles and practices.

Natural soils develop from the physical and chemical weathering of rocks and minerals over a geologic timescale. The properties of a given soil are determined by the combined effects of the main soil-forming factors: parent material (rock), the local climate, endemic biota, topography, and time [17]. The unique combination of soil-forming factors yield many different types of soil that can be broadly classified based on physical and chemical characteristics. Although the properties of individual soils vary, all natural soils exhibit characteristic layers along a line extending from the soil surface toward the center of the earth. These soil layers are referred to as horizons and together comprise the soil profile. Urban soils are extensively disturbed, typically compacted, and frequently developed on composite materials derived from exogenous sources. Composite materials—sand, silt, clay, and/or gravel—used to achieve the desired structural properties of urban soils may also contain construction debris, waste, or other technogenic materials [25]. The excavation, mixing, filling, compaction, and/or supplementation of soils in urban environments with exogenous geologic or technogenic materials yield a soil environment enormously modified from natural conditions. Concerns regarding the suitability of urban soils for green infrastructure and the need for specific technical characteristics for stormwater runoff infiltration have led to growing interest in engineered soils for urban green-gray stormwater management systems.

2.3 Engineered Soils and Filter Media

Material selection for engineered soils or reactive filter media should be based upon applicable local water quality regulations. Geomaterials underlying permeable pavements and/or green infrastructure provide both water storage and filtration capacity. To date, the potential of reactive geomaterials for stormwater treatment has not been fully exploited. Traditionally, subbase geomaterials for use in urban infrastructure are selected based on structural properties and their capacity for pollutant removal via sorption processes is not considered in detail. Geomaterials associated with gray infrastructure in particular are conventionally regarded as effectively inert with pollutant removal capacity limited to filtration of particulate materials.

Engineered stormwater management systems containing reactive geomedia can improve stormwater quality prior to water infiltration, uptake by collection systems, or discharge to receiving waterbodies. This, in turn, reduces pollutant loads to aquatic environments, thereby reducing eco-pressure on stressed water resources and associated ecosystems. In addition to returning cleaner freshwater to ecosystems, stormwater purification also facilitates its use as an alternative to potable water supplies and contributes to long-term water resource sustainability.

The mobility of pollutants in water can be decreased by filtering through a chemically reactive solid material with a large number of surface sorption sites, which creates a chemical environment favorable to sorption, precipitation or transformation of the element. In natural soil, most trace elements have low mobility because they adsorb strongly to soil minerals and organic matter or form insoluble precipitates. Urban stormwater filtration schemes build upon knowledge of ion interactions with mineral and organic phases in soil to design systems with the capacity to effectively remove elements from solution. Optimal pollutant mitigation by engineered soils and filter media requires a detailed understanding of existing local water quality concerns and pertinent water quality regulations and/or water quality objectives.

2.3.1   Selection of Filter Materials

Engineered soils and filter media (i.e., filter materials) used in stormwater runoff filtration are considered effective when local water quality objectives are achieved. Pollutants may be removed from stormwater runoff during infiltration through a combination of physical entrapment and sorption and/or precipitation reactions (Fig. 2.7). Stormwater pollutant adsorption and precipitation to filter material surfaces occur along a continuum, with the relative contribution of surface precipitation to pollutant immobilization increasing as the concentration of pollutant increases. Pollutant ions can interact with the solid surface through the formation of an inner-sphere or outer-sphere complex depending upon whether a chemical bond is formed. Adsorption reactions wherein the ion retains its hydration shell, and intervening water molecules remain between the ion and the material surface, are referred to as outer-sphere complexation. An outer-sphere complex does not involve a specific chemical bond between the solid surface and the adsorbed ion. Outer-sphere complexation is solely the result of electrostatic attraction between the ion and the solid surface, and ions associated with the solid surface through outer-sphere complexation are exchangeable. In contrast, ions adsorbed to a surface via inner-sphere complexation, wherein the ion adsorbs directly to the solid surface with no

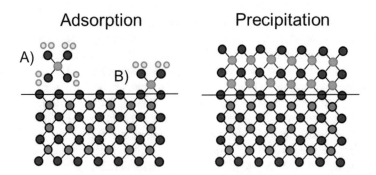

Fig. 2.7 Pollutant (pink ion) adsorption to and precipitation on a solid surface. Adsorption can occur through electrostatic attraction or outer-sphere complexation (A, non-specific) or through the formation of chemical bonds/inner-sphere complexation (B, specific) to the mineral surface

intervening water molecules, are far less exchangeable and the complexes formed are more durable. Formation of an inner-sphere complex indicates that the solid surface has a high affinity for the respective ion; inner-sphere complexes form when an ion adsorbs to a specific site on the surface, and they involve some degree of covalent bonding between the adsorbed ion and the solid surface.

Important characteristics of effective filter materials include both high specific surface area and surface reactivity. Porous or expandable solid filter materials have greater surface area per unit mass compared with nonporous/nonexpandable materials. Materials commonly used in passive or semi-passive filter systems to treat polluted water include alkalinity-producing materials such as limestone or calcined magnesia, reactive oxide minerals—typically iron, manganese, or aluminum oxides—and/or highly porous materials such as zeolite, activated carbon or biochar, expanded clay aggregate, or peat-based materials. Reactive materials may comprise as little as 2–10 wt.% of a filter system, with the bulk of the filter made up of locally available, nominally chemically inert materials, i.e., sand and gravel.

Studies of filter materials for removal of common metal and nutrient pollutants in stormwater runoff have shown that lead and some copper in influent stormwater were most likely removed through both physical trapping and chemical adsorption and/or precipitation reactions [36]. In contrast, zinc removal most likely occurred only through surface adsorption and precipitation. The study results further indicated that the availability of iron, manganese, and aluminum oxide surface functional groups on solid filter materials was particularly important with respect to zinc retention by the filter media. Phosphorus present as phosphate (PO_4^{3-}) in stormwater runoff was removed both by adsorption to oxide mineral surfaces and by the formation of calcium phosphate mineral precipitates. Thus, both the iron oxide mineral content and the content of available calcium in filter materials had a significant impact on phosphorus removal from stormwater. This study, which utilized a suite of metal and nutrient pollutants commonly identified in urban stormwater runoff, highlights the complexity of physical-chemical interactions among stormwater runoff constituents and filter materials.

Physicochemical interactions between pollutants common in stormwater and a wide range of potential solid filter materials are well documented in the scientific literature. Nevertheless, preliminary testing of pollutant removal from local stormwater runoff using possible filter materials integrated with local bulk materials in a configuration analogous to the planned filter design is recommended prior to full-scale deployment to ensure appropriate alignment between local water quality issues and the selected filter materials and structure design. A combination of laboratory- or pilot-scale testing and system modelling is ideal for optimizing filter design. Hydrologic modelling which accounts for stormwater pollutant differences based on runoff source areas can be employed to identify site-specific pollution challenges and solution constraints. In combination with appropriate laboratory testing and geochemical modelling, optimal filter solutions for site-specific stormwater runoff quality issues can be designed.

2.3.2 General Principles for Design of Engineered Filtration Systems

Batch tests are commonly used to estimate the capacity of solid materials to remove pollutants from solution. It is important to understand that specific pollutant removal capacities determined for individual materials or combined mixtures by batch testing are concentration-dependent. The net removal of a given pollutant represents its maximum potential removal at the concentration tested and in a static system without constant replenishment of solution. Partitioning of the pollutant between the solid material and the aqueous phase (K_d) represents both the affinity of the solid surface for the pollutant and the thermodynamic equilibrium between adsorbed (surface-associated) and free (in solution) pollutant ions. Thus, material testing using a relatively concentrated solution, i.e., an aqueous solution representative of metalliferous mine process water, will yield greater observed sorption of each pollutant, or higher sorption capacities, than would be determined for the same materials in a relatively dilute solution such as stormwater. For this reason, filter material sorption capacities should be estimated using a solution containing a mixture of pollutants at concentrations representative of the water to be treated.

The most important parameter affecting the behavior of metals and phosphorus in a filter system is the pH because it affects both solution and mineral surface chemistries. Metal adsorption to a solid material is typically low at acid pH and increases as solution pH approaches neutrality. The chemical form of a pollutant in solution, or its chemical speciation, is key to its behavior. Both metallic and nonmetallic pollutants may occur in different oxidation states and in soluble complexes with different organic or inorganic ions or molecules. In the absence of ligands capable of forming soluble complexes with metals (i.e., HCO_3^-, CO_3^{2-}, OH^-, organic acids), metals can be almost entirely removed from water at neutral to alkaline pH via adsorption to solid filter materials [3].

Hydraulic retention time (HRT), which is effectively the time during which infiltrating water remains in contact with the solid phase, is another key parameter for consideration in the selection of filter materials and filtration structure design. The HRT determines the length of time permitted for reaction between the solid filter material and pollutants in infiltrating water. In material testing, observed HRT variation between different test materials is due to differences in hydraulic conductivity among the respective materials or how easily water is able to pass through each filter material. The hydraulic conductivity is, in turn, dependent upon factors such as the filter material's grain size, the size and shape of pore spaces within the solid matrix, total porosity of the solid matrix, and the total surface area of the solid material per unit mass (e.g., the specific surface area). An ideal HRT allows sufficient time for optimal pollutant removal through interaction with solid filter materials and is dependent upon characteristics of both the stormwater runoff and the selected filter material(s). The HRT is typically determined in laboratory testing prior to filtration system design to facilitate appropriate system scaling for effective pollutant removal.

Column tests are commonly employed to evaluate the performance and effective lifetime of filter materials. In flow-through test systems, filter materials may be present as a single solid phase or as a mixture with nominally nonreactive materials. In the example shown in Fig. 2.8, filter materials were tested in a column design akin to planned full-scale deployment with the reactive filter material sandwiched between layers of nominally nonreactive aggregate.

The purpose of the aggregate layers above and below test filter materials shown in Fig. 2.8 was to slow the flow of influent stormwater in highly porous materials with low hydraulic retention time and to filter particulate materials [37]. Column testing is an important proxy for field-scale studies as the laboratory environment offers cost-effective control over system operation, monitoring, and sample collection compared with in situ experiments. Two key parameters that are typically calculated from column tests of filter materials are the breakthrough capacity (q_b) and saturation capacity (q_s) of the material in the tested configuration. The breakthrough capacity is defined as the sorption capacity of the filter material when a particular pollutant in the column effluent reaches a predefined concentration, for example, a regulatory limit for a given pollutant. The saturation capacity is reached when the filter material can no longer retain any additional quantity of a given pollutant and the concentration of that pollutant in column effluent is equal to that in the influent water. Column tests are most informative when filter materials are tested in a configuration representative of planned use and where the influent water and dosing regimen (e.g., continuous or periodic flow) closely resemble expected field conditions. In practice, spatial and/or financial constraints or compliance with specific material testing standard protocols dictate that column testing may be conducted at a range of scales from microscale (e.g., columns a few centimeters to tens of centimeters in length) to full-scale (e.g., columns typically 1 meter or more in length) prior to pilot-scale field study.

A further consideration related to HRT is oxidative-reductive potential (ORP). Together, pH and ORP are considered the master environmental variables that control pollutant speciation and behavior. Oxidation-reduction, or redox, reactions most commonly occur under waterlogged conditions where the rate of O_2 consumption

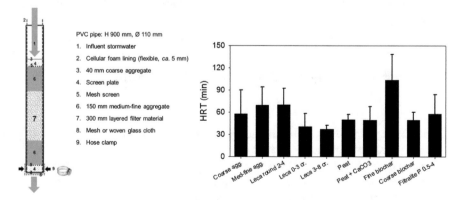

Fig. 2.8 Example experimental column design (left) and hydraulic retention time (HRT; right) in laboratory column trials. Error bars represent one standard deviation from the mean of triplicate measurements. (Reproduced with permission from Wendling et al. [37])

exceeds the rate of O_2 diffusion. Redox reactions are particularly relevant to nitrogen and iron/manganese dynamics but also affect a range of other potential pollutants. Rainfall is oxygenated because it is in equilibrium with atmospheric gases. Stormwater runoff can be viewed as a thin sheet of water moving over the surface of the landscape that can gradually become deoxygenated when collected in a stagnant system as a result of microbial or chemical activity that consumes O_2. Deoxygenation of stormwater is generally undesirable because discharge of deoxygenated water can have deleterious effects on aquatic ecosystems; however, in some instances treatment modules may be specifically designed to promote chemically reducing conditions in order to attenuate pollutants not readily removed in oxidized form. Nitrate (NO_3^-) is an example of a common stormwater pollutant usually removed by manipulation of redox conditions. In a stormwater treatment train, a final "polishing" step may include a module designed to induce chemically reducing conditions in order to remove nitrate via denitrification. Constructed wetlands are frequently used to collect stormwater. In deeper areas, where oxygen ingress is limited, denitrifying bacteria can use nitrate rather than oxygen in metabolic processes, thereby removing nitrate from the water and generating nitrogen gas.

Selection of reactive filter materials for a specific application should be based upon the characteristics of the influent stormwater. Thus, an understanding of both the chemical characteristics of stormwater runoff and the anticipated hydraulic loading to the stormwater filtration system is required for effective stormwater filter design. The expected duration of the material trapping functionality should further be considered in light of anticipated maintenance actions. Note that the capacity of a given filter system is largely determined by filter design and sizing relative to the pollutant load in influent stormwater runoff. An appropriately designed and properly scaled stormwater filter system can be expected to maintain pollutant removal functionality for at least 10–15 years. In general, a filtration system that utilizes decomposable organic material as a reactive phase can be expected to have a shorter lifetime relative to a filtration system comprised solely of inorganic materials with an equivalent pollutant retention capacity. A pre-treatment module to reduce TSS load to the filter can substantially lengthen filter system longevity by minimizing filter clogging.

2.4 Holistic Management of Stormwater Quantity and Quality

Systems of green infrastructure often utilize engineered infiltration or subsurface filtration media to optimize hydraulic conductivity/maximize water infiltration, filter particulate pollutants, or provide growth media for microbial communities. In general, urban green infrastructure has largely focused on filtration for stormwater *quantity* to minimize flooding rather than water *quality* for pollution control. Integrated engineered stormwater infiltration systems containing reactive geomedia can improve stormwater quality prior to water use, infiltration, uptake by collection systems, or discharge to receiving waterbodies. With improved gray infrastructure designs based on well-understood, quantifiable mechanisms of contaminant removal, engineered infiltration within integrated blue-green-gray systems of urban stormwater management infrastructure has the potential to provide a reliable holistic stormwater management solution, yielding a water resource that is safe for aquifer recharge, or to support healthy urban stream ecosystems. Stormwater purification facilitates its use in urban agricultural applications as an alternative to potable water supplies and contributes to long-term water resource sustainability.

Protection or restoration of both freshwater and marine receiving waters to meet water quality objectives is increasingly accomplished using decentralized engineered water management structures to trap, infiltrate, and/or harvest stormwater. These engineered structures are typically comprised of green infrastructure solutions or green-gray systems of infrastructure. Numerous studies have shown that vegetated stormwater filters can attenuate biochemical oxygen demand (BOD), suspended solids, nutrients, pesticides, and polycyclic aromatic hydrocarbons (PAHs) in stormwater runoff [2, 21, 41]. Examples of stormwater quantity and quality management systems include:

- Retention basins to store and slowly release or evaporate stormwater
- Engineered wetlands (Fig. 2.9) to attenuate contaminants in stormwater prior to discharge to natural ecosystems
- Permeable pavements, buffer strips, and biofiltration swales to remove suspended solids and encourage groundwater recharge
- Rain gardens, green roofs, and other bioretention systems which use vegetation to remove particulate and soluble contaminants, take up nutrients, and promote evapotranspiration
- Rain tanks and cisterns for capture and use of stormwater for localized irrigation or other non-potable applications

Engineered stormwater management solutions are broadly based on design elements with the following primary functions:

- Detention and infiltration
- Conveyance
- Pollution prevention
- Evapotranspiration
- Rainwater capture and use

These primary stormwater management functions may be accomplished using one or more modes of action, including (bio)filtration; retention (permanent pool); detention (temporary pool); infiltration; evaporation/evapotranspiration; biological, chemical, or physical pollutant removal or transformation; and storage for later use. Engineered stormwater solutions commonly employed in urban areas and their primary modes of action are summarized in Table 2.2.

The effectiveness of stormwater management solutions employing filtration and/or infiltration as a primary mode of action may be enhanced by the use of reactive filter media to further increase pollutant removal. Existing engineered and

Fig. 2.9 The Eerolanpuro engineered wetland in Jyväskylä, Finland, attenuates pollutants in stormwater runoff before discharge to nearby Lake Tuomiojärvi, which serves as a source of raw water for the city

nature-based stormwater management solutions provide the means to systematically address urban water management while highlighting the need for effective integration of multidisciplinary solutions and targeted future technological innovation. Integration of engineered blue-green-gray stormwater solutions into urban planning should be a joint effort between urban infrastructure owners (cities), urban planners/designers, environmental specialists, water management experts, material experts, and landscape designers familiar with vegetation.

2.5   Summary

Increased flooding is frequently cited as one of the most serious consequences of climate change for urban areas. Conventional water management via blocking and channeling water to rapidly convey stormwater away from inhabited areas provides a stormwater management solution under predictable, mild climatic conditions. The increasingly variable and extreme weather events resulting from climate change may be more effectively managed using decentralized water management structures to trap, infiltrate, and/or harvest stormwater. The integration of blue-green and gray urban infrastructure for stormwater management or coupling of centralized-decentralized urban water management systems can enhance urban areas' resilience to extreme weather events and climate change. A number of different decentralized stormwater management solutions are available; understanding how each component of a stormwater management system functions, e.g., its primary mode of

Table 2.2 Engineered solutions for urban stormwater management and their primary mode(s) of action

Primary class[a]	Engineered solution	Description and function
I	Public green space	Multifunctional public open space characterized by natural vegetation and permeable surfaces
R	Retention pond/wet detention pond	Pond wherein incoming stormwater replaces pond water; increases catchment water retention capacity, slows stormwater flow, and facilitates particulate settling. Storage capacity must be scaled based on local hydrology
D	(Dry) detention pond	Normally dry basin associated with a watercourse, usually equipped with spillway, designed to capture and slow stormwater runoff, reduce peak flows, and allow particulates to settle
I, D	Infiltration basin	Shallow, flat earthen depression to capture and store runoff until it infiltrates into surrounding soil; function to increase water storage capacity, slow runoff, filter particulate pollutants, and increase infiltration
I, B, E	Bioretention basin/bioretention cell	Landscaped shallow depression with associated vegetated filter or buffer strip; shallow ponding area is underlain by mulch, engineered soils and/or sand and an underdrain system. Harness natural biological and chemical processes to attenuate pollutants in runoff; function to slow runoff, filter pollutants, increase infiltration, and remove excess water to drain system
F, B	Sand filter	Constructed basin consisting of vegetated slopes surrounding a sand bed overlying an underdrain system
F, B, D, P	Surface stormwater wetland/marsh	Shallow, vegetated wet retention pond incorporating physical, chemical, and biological mechanisms of pollutant removal from influent water; increase system water retention capacity and slow stormwater flow
F, B, D, P	Subsurface wetland/filtration system	Constructed wetland with water level maintained below ground surface, planted with emergent aquatic vegetation and typically comprised of a gravel bed overlying a drainage system. Specifically designed for vertical or horizontal subsurface flow; increase water storage capacity/reduce peak flow, slow runoff, filter particulates
F, I	Riparian buffer zone	Vegetated linear area of land adjacent to a watercourse. Function to slow runoff and reduce flooding/increase infiltration, stabilize stream banks and reduce erosion, and filter particulate materials
I, B	Rain garden	Landscaped area with plants that can survive soil saturation; function to collect and slow stormwater runoff and increase water infiltration. Can be designed as household-sized bioretention basin
I, F	Vegetated filter strip	Gently sloping linear vegetated strips adjacent to impervious surfaces; reduce velocity of stormwater flow, filter particulates, and increase infiltration

(continued)

Table 2.2 (continued)

Primary class[a]	Engineered solution	Description and function
I, D, F	Wet/dry grassed swale, with or without check dams	Shallow open trapezoidal or parabolic open channel vegetated with flood-tolerant, erosion-resistant grasses; function to convey stormwater at a controlled rate, filter particulate pollutants, and enhance infiltration. Check dams' enhanced flow retardation and water retention capacity
I, D, F	Infiltration trench	Shallow excavated area filled with rock to create reservoir for runoff in void spaces between rocks. Runoff is captured and stored within the trench until it infiltrates into surrounding soil
I, F, E	Permeable pavement/pervious pavement	Permeable concrete or asphalt surfaces or permeable modular blocks, allow infiltration of stormwater to underlying soil. Mitigate flooding by decreasing runoff, increasing water storage capacity and infiltration, and filter particulates. Frequently combined with engineered soil infiltration bed and/or drain system
D, E, F	Green roof/green façade	Vegetated roof or wall cover consisting of layered membrane system, growth substrate, and living plants. Reduce runoff— highly variable due to differences in structure, growth media, and plant species/cover
S	Underground storage tank or cistern	Systems designed for capture and underground storage of stormwater runoff from impervious surfaces. Minimal water quality benefit in the absence of a coincident filtration system
S	Rain barrel or rain tank	Capture and store runoff from building roof, typically associated with irrigation storage. Function to increase water storage capacity/decrease flooding. Slow release of water between rain events reduces runoff and increases infiltration
P	Oil and grease separator/water quality inlet	Remove sediments, oil, and grease from roadway and parking lot runoff prior to discharge to storm drains

[a]*B* biofiltration, *D* peak volume reduction via detention (temporary pool), *E* evaporation/evapotranspiration, *F* filtration, *I* infiltration, *P* pollutant removal or transformation, *R* peak volume reduction via retention (permanent pool), *S* storage, typically for later use

action, facilitates the engineering design of coupled centralized-decentralized stormwater management systems for flood management.

The use of fit-for-purpose reactive filter media in decentralized green-blue stormwater management infrastructure can provide additional benefit in terms of stormwater quality improvements. Enhanced stormwater management solutions that remove pollutants from stormwater runoff, thus reducing pollutant loading to receiving waterbodies, are increasingly necessary to meet regulatory guidelines established to protect the ecological integrity of aquatic systems. The removal of stormwater contaminants associated with particulates occurs via physical trapping; however, stormwater contaminants not associated with TSS are unlikely to be attenuated by nonreactive filter media such as quartz sand. The use of selected filter media with the capacity to remove target contaminants from stormwater within green-blue stormwater management systems can increase the effectiveness of contaminant removal.

The integration of distributed and centralized, green-blue, and conventional gray infrastructure-based stormwater management solutions can also minimize the risks related to water scarcity in urban areas. Water scarcity currently affects at least 11% of the European population and 17% of the EU territory [9], with continuously increasing demand further straining existing freshwater resources. Coastal urban areas are especially vulnerable to freshwater scarcity due to the increased susceptibility of coastal groundwater to salinization. Where stormwater is captured and treated to a sufficient standard, either by green-blue decentralized treatment systems or by an integrated centralized-decentralized green-blue-gray stormwater treatment system, the treated stormwater can be used in non-potable water reuse schemes. For example, treated stormwater may be directly used to irrigate public green space, injected to recharge a depleted aquifer, used to augment the water supply in a drinking water reservoir, or discharged to an urban stream to maintain environmental flows during dry periods.

Improved urban stormwater management systems are an important yet relatively simple and economic component of a holistic urban water resource management scheme. With improved green-gray/blue-green-gray infrastructure designs based on well-understood, quantifiable mechanisms of contaminant removal, engineered infiltration within integrated decentralized urban stormwater management systems has the potential to substantially contribute to a reliable holistic urban stormwater management solution, yielding a water resource that is safe for aquifer recharge or to support healthy urban stream ecosystems.

References

1. Al-Rubaei AM, Stenglein AL, Viklander M, Blecken G-T (2013) Long-term hydraulic performance of porous asphalt pavements in northern Sweden. J Irrig Drain Eng 139:499–505
2. Barbosa AE, Jacobsen HT (1999) Highway runoff and potential for removal of heavy metals in an infiltration pond in Portugal. Sci Total Environ 235:151–159
3. Bhatia M, Goyal D (2014) Analyzing remediation potential of wastewater through wetland plants: a review. Environ Prog Sustain Energy 33(1):9–27
4. Bradl HB (2004) Adsorption of heavy metal ions on soils and soils constituents. J Colloid Interface Sci 277:1–18
5. Chessman BC, Hutton PE, Burch JM (1992) Limiting nutrients for periphyton growth in sub-alpine, forest, agricultural and urban streams. Freshw Biol 28:349–361
6. Comber SDW, Merrington G, Sturdy L, Delbeke K, Van Assche F (2008) Copper and zinc water quality standards under the EU Water Framework Directive: the use of a tiered approach to estimate the levels of failure. Sci Total Environ 403:12–22
7. Cooley LA Jr, Brumfield JW, Mallick RB, Mogawer WS, Partl MN, Poulikakos LD, Hicks G (2009) Construction and maintenance practices for permeable friction courses. NCHRP Report 640. National Cooperative Highway Research Program Transportation Research Board, Washington, DC, 133 pp

8. European Commission (2000) Directive 2000/60/EC of the European Parliament and of the Council of 23 October 2000 establishing a framework for community action in the field of water policy water framework directive. http://eur-lex.europa.eu/legal-content/EN/TXT/?uri= CELEX:02000L0060-20140101

9. European Commission (2008) Directive 2008/105/EC of the European Parliament and of the Council on environmental quality standards in the field of water policy. http://eur-lex.europa.eu/legal-content/EN/TXT/?uri=CELEX:32008L0105

10. European Commission (2010) Water scarcity and drought in the European Union. http://ec.europa.eu/environment/pubs/pdf/factsheets/water_scarcity.pdf

11. European Commission (2012) Guidelines on best practice to limit, mitigate or compensate soil sealing. European Union SWD(2012) 101

12. European Environment Agency (2012) Climate change, impacts and vulnerability in Europe 2012. EEA Report No 12/2012. http://www.eea.europa.eu/media/publications/

13. European Environment Agency (2012) Water resources in Europe in the context of vulnerability. EEA Report No 11/2012. http://www.eea.europa.eu/media/publications/

14. Fang X, Zech WC, Logan CP (2015) Stormwater field evaluation and its challenges of a sediment basin with skimmer and baffles at a highway construction site. Water 7:3407–3430

15. Garn HS (2002) Effects of lawn fertilizer on nutrient concentration in runoff from lakeshore lawns, Lauderdale Lakes, Wisconsin. USGS Water Resources Investigations Report 02-4130

16. Gill LW, Ring P, Higgins NMP, Johnston PM (2014) Accumulation of heavy metals in a constructed wetland treating road runoff. Ecol Eng 70:133–139

17. Granier L, Chevreuil M, Carru A-M, Létolle R (1990) Urban runoff pollution by organochlorines (polychlorinated biphenyls and lindane) and heavy metals (lead, zinc and chromium). Chemosphere 21(9):1101–1107

18. Jenny H (1941) Factors of soil formation: a system of quantitative pedology. McGraw-Hill, New York

19. Kevern JT, Schaefer VR, Wang K, Suleiman MT (2008) Pervious concrete mixtures for improved freeze-thaw durability. J ASTM Int 5(2):1–12

20. Kuosa H, Niemeläinen E, Loimula K (2013) Pervious pavement systems and materials. State-of-the-art. Research Report VTT-R-08222-13. VTT Technical Research Centre Ltd, Espoo, Finland, 95 pp. http://www.vtt.fi/files/sites/class/D2_1_CLASS_WP2_SOTA_Permeable_Pavement_systems_and_materials.pdf

21. Kuosa H, Loimula K, Niemeläinen E (2014) Vettä läpäisevät pinnoitteet ja rakenteet – Materiaalikehitys ja simulointitestaus. Tutkumusraportti VTT-R-05001-14. VTT Technical Research Centre Ltd, Espoo, Finland, 118 pp. [Finnish]. http://www.vtt.fi/files/sites/class/D2_5_CLASS_WP2_D2_Lab_and_Simulation_Results.pdf

22. Ladislas S, Gérente C, Chazarenc F, Brisson J, Andrès Y (2015) Floating treatment wetlands for heavy metal removal in highway stormwater ponds. Ecol Eng 80:85–91

23. Line DE, Shaffer MB, Blackwell JD (2011) Sediment export from a highway construction site in Central North Carolina. Trans ASABE 54(1):105–111

24. Liu Q, Cao D (2009) Research on material composition and performance of porous asphalt pavement. J Mater Civ Eng 21(4):135–140

25. Lundy L, Ellis JB, Revitt DM (2012) Risk prioritisation of stormwater pollutant sources. Water Res 46:6589–6600

26. Meuser H (2010) Anthropogenic soils. In: Contaminated urban soils. Environmental pollution, vol 18. Springer, Dordrecht, pp 121–193

27. Rabalais NN, Diaz RJ, Levin LA, Turner RE, Gilbert D, Zhang J (2010) Dynamics and distribution of natural and human-caused hypoxia. Biogeosciences 7:585–619

28. Rabatyagov SS, Kling CL, Gassman PW, Rabalais NN, Turner RE (2014) The economics of dead zones: causes, impacts, policy challenges, and a model of the Gulf of Mexico hypoxic zone. Rev Environ Econ Policy 8:58–79

29. Roads and Maritime Services (2017) Water sensitive urban design guidelines. Applying water sensitive urban design to NSW transport projects. Roads and Maritime Services Centre for

Urban Design and Environmental Land Management Section, and the Infrastructure and Services Division of Transport for NSW, Sydney, 42 pp

30. Roseen RM, Ballestero TP, Houle JJ, Briggs JF, Houle KM (2012) Water quality and hydrologic performance of a porous asphalt pavement as a storm-water treatment strategy in a cold climate. J Environ Eng 138(1):81–89

31. Steuer J, Selbig W, Hornewer N, Prey J (1997) Sources of contamination in an urban basin in Marquette, Michigan and an analysis of concentrations, loads, and data quality. U.S. Geological Survey Water-Resources Investigations Report 97-4242

32. Tennis PD, Leming ML, Akers DJ (2004) Pervious concrete pavements. PCA Engineering Bulletin EB302, Portland Cement Association, 32 pp

33. Turner RE, Rabalais NN, Justić D (2012) Predicting summer hypoxia in the northern Gulf of Mexico: redux. Mar Pollut Bull 64:319–324

34. U.S. Environmental Protection Agency (1993) Guidance specifying management measures for sources of nonpoint source pollution in coastal waters United States Environmental Protection Agency #840-B-92-002. Washington, DC: USEPA Office of Water

35. U.S. Environmental Protection Agency (2002) Summary table for the nutrient criteria documents. http://www2.epa.gov/sites/production/files/2014-08/documents/criteria-nutrient-ecoregions-sumtable.pdf

36. United Nations, Department of Economic and Social Affairs, Population Division (2014) World urbanization prospects: The 2014 revision. United Nations, Rome

37. Wendling L, Loimula K, Kuosa H, Korkealaakso J, Iiti H, Holt E (2017) StormFilter Material Testing Summary Report. Performance of stormwater filtration systems. VTT Research Report VTT-R-05545-17. VTT Technical Research Centre of Finland, Espoo, 50 pp. http://www.vtt.fi/sites/stormfilter/Documents/VTT_R_05545_17.pdf

38. Wendling L, Loimula K, Kuosa H, Korkealaakso J, Iiti H, Holt E (2017) StormFilter Material Testing Summary Report. Localized performance of bio- and mineral-based filtration material components. VTT Research Report VTT-R-01757-17. VTT Technical Research Centre of Finland, Espoo, 55 pp. http://www.vtt.fi/sites/stormfilter/Documents/VTT_R_01757_17_1708.pdf

39. Wernick BBG, Cook KE, Schreier H (1998) Land use and streamwater nitrate-N dynamics in an urban-rural fringe watershed. J Am Water Resour Assoc 34(3):639–650

40. Westerlund C, Viklander M, Bäckström M (2003) Seasonal variations in road runoff quality in Luleå, Sweden. Water Sci Technol 48(9):93–101

41. Wolman MG, Schick AP (1967) Effects of construction on fluvial sediment; urban and suburban areas of Maryland. Water Resour Res 3(2):451–462

42. Zhang K, Randelovic A, Page D, McCarthy DT, Deletic A (2014) The validation of stormwater biofilters for micropollutant removal using in situ challenge tests. Ecol Eng 67:1–10

Laura A. Wendling is a Senior Scientist in the Infrastructure Health research group at VTT Technical Research Centre of Finland Ltd. She earned her Ph.D. in Soil Science from Washington State University (Pullman, WA, 2004), where she was a National Science Foundation IGERT Fellow. She is an experienced geoscientist with particular expertise in physicochemical interactions at the solid-water interface. Dr. Wendling's current research emphasis is on the development and application of novel technologies for sustainable water and soil management. She has made advances in the innovative application of mineral-based by-products as nutrient and metal sorbents for wastewater treatment, providing an effective, affordable alternative to highly mechanized, energy-intensive technologies. Dr. Wendling has contributed her expertise in soil science, water quality, environmental chemistry, and wetland science to numerous collaborative and multinational research projects in the USA, Australia, China, and Finland. Her work has resulted in more than 100 scientific publications. She is a member of international networks involved with water and soil management issues, such as the NEREUS COST Action ES1403 *New and emerging challenges and opportunities in wastewater reuse*. Dr. Wendling serves as a trusted advisor to national and regional governments, municipalities, and business and industry clients with respect to water- and soil-related issues.

Dr. Wendling grew up in an outdoorsy family with parents who encouraged her avid interest in the natural world. Rachel Carson was her first inspiration, not only because she made significant scientific discoveries but because she remained steadfast in the face of harsh criticism. Rachel Carson, Marie Curie, Rosalind Franklin, and Theo Colburn are among the women who continue to inspire her. The public debate around oil drilling in the Arctic National Wildlife Refuge in the mid-1980s sparked Dr. Wendling's interest in a career where she could use science and engineering to develop solutions for environmental problems. Dr. Wendling studied biology as an undergraduate, and soil science as a graduate discipline. Soil science fascinates her because it integrates quantitative analyses and fundamental mechanistic knowledge across a broad range of subdisciplines (physics, chemistry, biology, fertility, and pedology).

The best career advice Dr. Wendling has received was from her father who told her to do something she really enjoyed because she would spend the best part of her life at work. Combined with advice from an academic advisor, who told her there are always jobs for those at the top of their field, Dr. Wendling charted her own career path. Hers is not a career described in glossy university brochures, and even now, it is sometimes difficult to describe her job to nonscientists. The journey has taken Dr. Wendling from the Pacific Northwest USA where she worked for the USDA-ARS to Australia, where Dr. Wendling spent nearly a decade working for the Commonwealth Scientific and Industrial Research Organisation. Dr. Wendling is now settled in Finland where she works to develop and test innovative technical solutions to contemporary environmental issues.

Dr. Erika E. Holt is a principal scientist within the Lifecycle Solutions business area at VTT Technical Research Centre of Finland Ltd. She has previously been the team leader for Infrastructure Health, with an emphasis on material-based technology solutions for water management. Erika holds a Ph.D. from the University of Washington (Seattle, 2001) in Civil and Environmental Engineering, with an emphasis on building material durability. She has worked at VTT in Finland since 1996, where her research areas often address utilization of sustainable materials and service life approaches to material performance. For the past 5 years, she has been coordinating large Finnish projects related to innovations for managing urban stormwater quality and quantity, e.g., the development of arctic durable pervious pavements. She is a member as the European Construction Technology Platform and RILEM (International Union of Laboratories and Experts in Construction Materials, Systems and Structures). Erika is responsible for project management, project development, client interactions, international networking, technical reviews, and strategic planning. She manages a portfolio of VTT's jointly funded (public) projects related to the Finnish and international nuclear sector (~250 VTT persons), which also has environmental safety priorities associated with groundwater and surface water characteristics. She has over 100 scientific publications and a strong liaison with industrial clients, including cities, material producers, and designers implementing water management best practices.

Dr. Holt originally studied geotechnical engineering, with respect to geology and tunnelling. Her father was a construction consultant, and Erika loved being outdoors and "building things" as a kid, from tree houses in the forest to rock dams across the creek in her backyard. She enjoyed camping at national parks, where she remembers visiting caverns, craters, and other geological wonders. Dr. Holt had inspiring teachers and mentors through school who encouraged her to pursue scientific studies and internships including engineering field work around the USA. During graduate school in Civil Engineering, Dr. Holt also had an emphasis on construction materials, which led her to Finland on a research grant to investigate durability of concrete in harsh environments. Dr. Holt continued her career in Finland, and she conducts research on sustainable and environmental practices for underground spaces. Her investigations have included urban stormwater management using engineered soils, permanent geological storage for safe containment of nuclear spent fuel, and the recycling of industrial by-products into construction materials.

Chapter 3
Improving Drinking Water Quality in Rural Communities in Mid-Western Nepal

Sara Marks and Rubika Shrestha

Abstract Achieving universal access to safe drinking water is a global challenge, especially in rural areas of low-income countries. In Nepal, most rural households have access to a protected drinking water source. However, for 75% of the rural population, these sources are impacted by fecal and chemical contaminants. This chapter describes 4 years of applied research on drinking water quality in Mid-Western Nepal, in collaboration with Helvetas Swiss Intercooperation and the REACH: Improving Water Security for the Poor program. The aim of this project was to improve access to safe drinking water for rural households served by the Helvetas Water Resources Management Programme. The field activities were organized into three phases: a baseline characterization of microbial quality at water collection points and household water storage containers for 505 households; an investigation of households' perceptions and practices regarding household water treatment; and controlled evaluation of a combined water safety intervention's impact on *E. coli* concentrations for five piped schemes. The interventions examined included solar-powered field laboratories, centralized data management, targeted infrastructure improvements, household filter promotion, a sanitation and hygiene behavior change campaign, and community-level orientation and training. By the end of the study period the share of taps and storage containers meeting the WHO guideline for microbial safety increased from 7% to 50% and from 17% to 53%, respectively. These findings indicate that a combination of tailored interventions can effectively reduce fecal contamination at the points of collection and consumption for piped supplies in remote rural communities.

S. Marks (✉)
Eawag, Swiss Federal Institute of Aquatic Science and Technology, Dübendorf, Switzerland
e-mail: sara.marks@eawag.ch

R. Shrestha
Helvetas Swiss Intercooperation Nepal, Kathmandu, Nepal
e-mail: Rubika.Shrestha@helvetas.org

© Springer Nature Switzerland AG 2020
D. J. O'Bannon (ed.), *Women in Water Quality*, Women in Engineering and Science, https://doi.org/10.1007/978-3-030-17819-2_3

3.1 Introduction

Considerable progress has been made in recent decades extending access to drinking water services globally. From 2000 to 2015, 2.6 billion people worldwide gained access to improved drinking water, defined as a source that is protected from outside contamination [13]. Still, half a million people worldwide died in 2012 due to preventable waterborne diseases [7] and sustainability of water infrastructure remains challenging [16, 17] [montgomery, marks].

In response to these issues, water sector professionals have adopted Sustainable Development Goal (SDG) Target 6.1, which aims "to deliver, by 2030, universal and equitable access to safe and affordable drinking water for all" [14]. Target 6.1 specifically addresses availability, affordability, and quality as core priorities in its definition of safely managed drinking water. Based on these standards, over a quarter of the global population currently lacks access to safe water, with most of the unserved living in rural areas of the least developed nations [14].

This chapter details 4 years of applied research on drinking water quality of piped schemes in remote rural communities in the Mid-Western region of Nepal [1, 2, 8]. The project was implemented by Helvetas Swiss Intercooperation's (hereafter referred to as Helvetas) Water Resources Management Programme (WARM-P), in collaboration with the Swiss Federal Institute of Aquatic Science and Technology (hereafter referred to as Eawag) and the REACH: Improving Water Security for the Poor program (an initiative of Oxford University and funded by the UK Government).

3.2 Background and Study Site

3.2.1 Drinking Water and Sanitation in Rural Nepal

Nepal was ranked at the poorest end of the UNDP Human Development Index in 2016, in the 144th position out of 188 countries [10]. Water scarcity is a common issue in the country [6, 9] that is exacerbated by ongoing climate change effects [15]. While coverage by protected drinking water sources (such as public taps or rooftop rainwater harvesting systems) is nearly 90% in rural Nepal, most of these sources are impacted by fecal and chemical contamination (http://washdata.org/). Access to sanitation is lower, with 45% of the rural population using an improved facility, such as a latrine or pour-flush toilet. In the hill areas of the Mid-Western region, where development indicators show some of the country's highest poverty rates [3], access to a basic or safely managed drinking water source is among the lowest nationwide at 69% (Fig. 3.1). Achieving SDG 6.1 is especially challenging in this setting, where barriers to implementing water supply treatment and monitoring include unreliable material supply chains, the high cost of laboratory equipment, and low or nonexistent access to electricity.

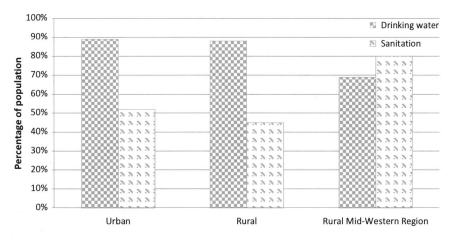

Fig. 3.1 Access to basic or safely managed drinking water and sanitation facilities in 2015 in Nepal (urban, rural, and rural Mid-Western region). According to the WHO/UNICEF Joint Monitoring Programme, access to *basic water* is defined as "drinking water from an improved source, provided collection time is not more than 30 minutes for a roundtrip including queuing," and access to *safely managed water* is defined as "drinking water from an improved source, located on premises, available when needed and free from fecal and priority chemical contamination." (source: http://washdata.org/)

3.2.2 Study Site and Objectives

Helvetas WARM-P aims to identify water resources and foster effective, equitable, and efficient use at the local and regional levels [4]. Since 2001, WARM-P has installed over 300 gravity-fed piped water systems across five districts (Achham, Dailekh, Kailali, Kalikot, and Jajarkot) (Fig. 3.2). Most of the systems provide intermittent water services with variable opening times and service durations throughout the year, depending on the season. Figure 3.3 shows the typical layout of a WARM-P water system, with a spring source that is connected to a reservoir tank by a distribution line, with water then flowing to the taps. Within the WARM-P service area, there is the added goal to improve drinking water quality through a demand-led approach for ensuring treatment and safe storage at the household level.

Helvetas partnered with Eawag and REACH to address the following research questions relevant to WARM-P:

- To what extent are community drinking water supplies impacted by fecal contamination?
- What are households' perceptions of their water services, and is there demand for safer drinking water?
- Can a combined water safety intervention that includes regular monitoring, system upgrades, and hygiene behavior change improve drinking water quality at the points of collection and consumption?

Terai
Hilly area
Map of Nepal

Fig. 3.2 Map of Nepal showing district boundaries and the service coverage area of the Helvetas Water Resources Management Programme (WARM-P) in red and orange. The hilly area is situated to the south of the Himalayan region and is characterized by natural vegetation and limited crop cultivation. The terai is a lowland region characterized by more intensive cultivation and dense population. (Figure credit: Helvetas Swiss Intercooperation)

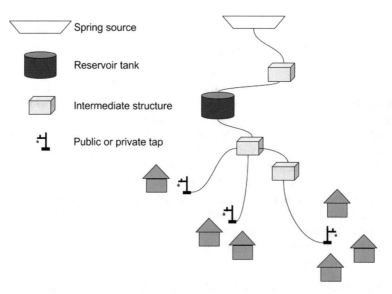

Spring source

Reservoir tank

Intermediate structure

Public or private tap

Fig. 3.3 A typical gravity-fed piped water scheme (or sub-scheme) in the WARM-P service area. Each sub-scheme makes use of independent water sources. (Figure credit: Dorian Tosi Robinson)

The study addressed these questions in three phases of field activities from 2014 to 2018, with the goal of ultimately implementing a risk management strategy in support of meeting SDG 6.1 for gravity-fed piped schemes across the WARM-P service area.

3.3   Research Phases

3.3.1   Phase 1: Assessing Microbial Quality of Piped Water Supplies

Objective To assess fecal contamination concentrations at community taps ($n = 100$) and within households' water storage containers ($n = 505$) for five communities served by gravity-fed piped systems.

Methods Water samples were collected in a sterile Whirl-Pak® bag from a water storage container in each household and transported within 6 h to temporary field laboratory stations centrally located within each community. A sample was also taken at each public tap in the five study communities. Samples were processed in the field laboratories using a modified membrane filtration set up with Hyserve compact dry plates to determine *E. coli* concentrations. Plates were placed in a solar-powered incubator at 35 ± 2 °C for 24 h before colony-forming units (CFU) were counted (Fig. 3.4).

Fig. 3.4 Photos of field equipment for microbial water quality testing in Nepal. (Photo credit: Dorian Tosi Robinson and Sara Marks)

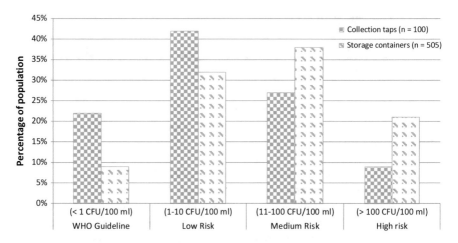

Fig. 3.5 Water quality at collection taps and within household storage containers

Results Most of the piped water systems operated for less than 24 h/day (intermittent supply). Nearly all samples taken from water storage containers (91%) had detectable *E. coli* in excess of the WHO guideline for microbial safety of drinking water [11], and 21% of stored water samples contained more than 100 CFU/100 mL, indicating a very high risk to health. At collection taps, less than one-quarter (22%) of the samples collected had no detectable *E. coli*, and 9% samples contained more than 100 CFU/100 mL (Fig. 3.5).

Water quality deteriorated from the tap to the storage container in 72% of households, indicating recontamination during transport and handling. Fecal bacteria concentrations at the tap were positively correlated with concentrations in the storage container it supplied (Spearman's ρ (283) = 0.25, $p < 0.001$). For water systems that households identified as providing water continuously (no service interruptions), microbial concentrations were significantly lower than for systems experiencing daily interruptions (Mann-Whitney U (196) = 3380, $p < 0.05$).

Summary Phase 1 of the project revealed that most tap samples did not meet international standards for drinking water safety. The vast majority of stored water samples were contaminated with fecal bacteria, indicating that water quality deteriorated between the point of collection and storage due to recontamination during transport and handling. In addition, about one in ten water samples taken at collection taps had highly elevated concentrations of *E. coli* (>100 CFU/100 mL), highlighting the inadequacy of infrastructure-centered definitions for an "improved" water sources.

3.3.2 Phase 2: Understanding Households' Drinking Water Perceptions and Practices

Objective To assess water users' perceptions regarding their drinking water supply, current household water treatment practices, and demand for improved drinking water safety.

Methods Six enumerators conducted semi-structured interviews with 512 households using tablets loaded with Open Data Kit (ODK) software. All enumerators were fluent in Nepali and were trained over the course of 1 week. Survey questions assessed households' water management practices, perceptions regarding their water services, socioeconomic status, and environmental factors. Each survey required 30–45 min to complete. Households were randomly selected for participation in the study; following a community mapping exercise, every second or third household (depending on community size) along a transect walk was selected for enrollment for a total of about 100 households per district. Participating households enrolled following oral informed consent, with the agreement that participation in the survey was voluntary, all questions were optional, and the information obtained would remain anonymous.

Results Half of the households interviewed had no educational background, while 23% went to primary school and 27% to secondary school or higher. Only 4% of households had adequate hand-washing facilities (defined as the presence of soap and water), and 41% of respondents reported washing their hands two times daily or less. Most households (87%) had a ventilated improved pit (VIP) latrine. In the past 3 days, 5% of children under age 5 had experienced diarrheal disease, and 19% had experienced acute respiratory illness. Nine out of every 10 households interviewed had access to an improved source for their main drinking water supply, with about half using a community tap from the piped water system. However, in Kailali, where the topography is less hilly, using tube wells was the standard practice. Most households (91%) reported using the same container for transport and storage. Most water schemes operated for less than 24 h/day (intermittent supply).

Households were generally aware of the main sources of fecal contamination of water (animal and human excreta). However, many households did not perceive risks to their own drinking water supply, with two-thirds reporting that it was either good or very good quality (Fig. 3.6) and 46% attributing no or little diarrhea risk to drinking untreated water. Nearly half of the households interviewed said drinking water directly from the source was safe, while only 11% said it was very risky. As such, most respondents (69%) said they did not treat their own drinking water and only 12% intended to treat in the future. Knowledge of different water treatment technologies was very limited at the time of the study; most respondents (70%) could not explain the proper steps for more than one treatment method, such as boiling, ceramic filtration, or chlorine disinfection.

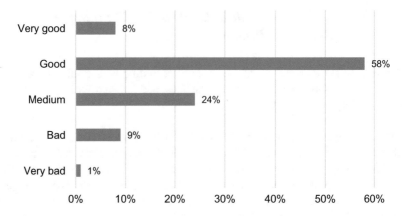

Fig. 3.6 Households' perception of drinking water quality ($n = 505$)

Household water treatment use was significantly and positively correlated with having access to treatment products locally, emotional factors regarding water treatment, having sufficient knowledge of treatment methods, a stated intention to treat, and believing it is important to treat water (Spearman's ρ, all $r > 0.3$, $p < 0.001$).

Summary The results of Phases 1 and 2 suggest a need for a risk-based approach to protecting and mitigating scheme- and household-level water quality. In particular, Phase 2 revealed that most households perceived their own drinking water to be of good quality and did not practice nor prioritize investments in treatment methods [5]. An effective behavior change campaign should target knowledge on personal risk and mitigation options, emotional factors, and the perception of personal vulnerability to highlight the importance of consuming safe water. An intervention should also support households to develop strategies for accessing safe water products and establishing local hand-washing facilities. Introduction of viable household water treatment options, in combination with these behavioral interventions, may improve water quality at the point of consumption.

3.3.3   Phase 3: Measuring the Impact of Water Safety Interventions on Microbial Water Quality

Objective To assess the effectiveness of a risk management strategy to improve drinking water quality at collection taps and within household water storage containers for five gravity-fed piped schemes [8].

Methods The study was conducted in the Dailekh district of the Mid-Western Development Region (Fig. 3.2). This district was selected as the study location because it is representative of the rural, hilly areas that comprise about one-third of

Nepal's total land area. Dailekh district offered the additional advantages of close proximity to the Helvetas WARM-P office in Surkhet and relatively convenient road access. In total, eight communities with gravity-fed piped drinking water schemes were selected for the study: five schemes where a combination of water safety interventions were provided and three control schemes where no interventions were provided.

Prior to the study, all eight communities had received a new piped water system with private or public taps, constructed by Helvetas between 2012 and 2016. Following construction in each community, Helvetas had established a water and sanitation users' committee, promoted improved household hygiene practices, promoted ceramic filters for household water treatment and safe storage, and trained a female community health volunteer and a village maintenance worker. The study interventions examined included field laboratories for microbial analysis, centralized data management, targeted water infrastructure improvements, chlorination of reservoir tanks, intensive household hygiene and filter promotion, and training of a community water safety task force and laboratory technicians (Table 3.1).

The following measures were assessed before and after implementation of the water safety interventions: community members' perceptions and behaviors regarding their drinking water ($n = 120$), the sanitary state of the water schemes ($n = 23$), and E. coli concentrations at collection taps ($n = 23$) and household storage containers ($n = 120$). The interventions described in Table 3.1 were implemented over an 8-month period between the two rounds of data collection (hereafter called "pre-intervention" and "post-intervention"). The methodologies for water sampling and household survey data collection are as described above in Phases 1 and 2, respectively.

Results Results showed a significant improvement in microbial water quality within households and at taps following implementation of the interventions (see Table 3.2). Within intervention schemes, the mean $\log_{10}$ E. coli concentration in

Table 3.1 Activities carried out within intervention and control communities during Phase 3

Activities carried out prior to the study in all communities
Constructed piped water system construction
Established water and sanitation users' committee
Delivered household hygiene messages
Promoted ceramic water filters
Trained community health volunteers and village maintenance workers
Activities carried out during the study in intervention communities only
Installed field laboratories
Established centralized data management
Infrastructure improvements, e.g., pipe replacements, tank repairs, intake filters, etc.
Chlorination of reservoir tanks
Delivered water quality test results along with intensive household hygiene and water safety messages
Trained water safety task force members and laboratory technicians

Table 3.2 *E. coli* concentrations at each sample location within the intervention and control communities, with bivariate comparisons of the mean *E. coli* contamination at pre- and post-intervention time points

Location	Before or after intervention	n	Percentage samples meeting WHO guidelines (%)	Median [CFU/100 mL]	Mean (SD) [$\log_{10}$(CFU/100 mL)]	Student's t-test
Intervention schemes						
Household stored water	Before	75	17	24	1.25 (1.00)	t = −5.645, df = 145, **p < 0.001**
	After	72	53	0	0.36 (0.92)	
Collection tap	Before	14	7	11	1.14 (0.79)	t = −4.086, df = 26, **p < 0.001**
	After	14	50	1	0.13 (0.49)	
Control schemes						
Household stored water	Before	45	20	8	1.01 (0.97)	t = −1.026, df = 86, p = 0.308
	After	43	23	4	0.80 (0.98)	
Collection tap	Before	9	0	38	1.54 (1.01)	t = −2.040, df = 16, p = 0.058
	After	9	11	3	0.65 (0.82)	

household storage containers was 1.25 CFU/100 mL at the pre-intervention time and 0.36 CFU/100 mL at the post-intervention time. Among intervention scheme taps, a reduction in the mean $\log_{10}$ *E. coli* concentration from 1.14 CFU/100 mL to 0.13 CFU/100 mL was observed. There was no significant difference in the average contamination levels observed at pre- and post-intervention for any of the sampling points in the control schemes.

Each sampling point was examined for compliance with WHO guidelines for drinking water safety (<1 CFU *E. coli*/100 mL). The results show that, within intervention schemes, the share of household stored water samples with no detectable *E. coli* increased significantly from 17% to 53% following implementation of the interventions (c2 (1, *n* = 147) = 24.01, *p* < 0.001). The increase in the tap samples from the intervention schemes that met the WHO guidelines was also significant, from 7% to 50% (c2 (1, *n* = 28) = 6.30, *p* = 0.03). In addition, all the tap samples taken post-intervention had less than 10 CFU *E. coli*/100 mL.

Summary Within the 5 intervention schemes, there was a statistically significant improvement in microbial water quality 8 months after implementation of the water safety measures. All taps within intervention schemes had a concentration of less than 10 CFU *E. coli*/100 mL at the end of the study period. These water quality improvements were driven by scheme-level chlorination, improved household hygiene practices, and the universal adoption of household water treatment. Implementation of this comprehensive risk management strategy in remote rural communities can support efforts toward achieving universal access to safely managed water.

3.4 Conclusions

The findings from 4 years of research in rural communities in Mid-Western Nepal reveal the importance of a comprehensive strategy for reducing contamination of piped water supplies. At baseline 90% of the households interviewed had access to an improved drinking water source, such as a tap or borehole. Most households understood the main sources of contamination and the importance of using a safe drinking water source. Most households also perceived their own water supply as safe to drink, and they therefore did not practice any treatment method. However, a sampling campaign revealed that two-thirds of taps and nine out of every ten household storage containers were contaminated with fecal bacteria. Thus, this research revealed a disconnect between households' perceptions and the actual microbial safety of their drinking water, resulting in a high risk to health due to exposure to waterborne pathogens. These findings pointed to a need for a campaign to increase hygiene education and awareness, as well as technical upgrades for the piped schemes across the WARM-P service area.

The final year of field activities in the WARM-P service area was dedicated to the implementation of a suite of water safety interventions, including pipe repairs, scheme-level chlorination, and intensive water safety and household hygiene promotion. The promotion strategy included door-to-door visits and community meetings to disseminate water quality test results and motivate treatment and safe storage practices. The evaluation of the impact of these interventions showed that such a risk-based approach resulted in measurable improvements in microbial water quality at taps and within household storage containers across the study area. For example, following intervention implementation, the share of taps and storage containers meeting the WHO guideline for drinking water safety increased from 7% to 50% and from 17% to 53%, respectively. In addition, by the end of the study period, all tap water samples contained less than 10 CFU $E.\ coli$/100 mL, a significant improvement from the pre-intervention levels observed among WARM-P piped systems. This applied research project was made possible by a long-term collaborative partnership between a development organization and academic institutions, with priority placed on stepwise learning processes and participatory involvement of all partners.

Acknowledgments This research was made possible by a team of interdisciplinary researchers, interns, and practitioners: (in alphabetical order) Madan Bhatta, Mohan Bhatta, Arnt Diener, Manuel Holzer, Jennfier Inauen, Bal Mukunda Kunwar, Laila Lüthi, Regula Meierhofer, Ariane Schertenleib, Ram Shrestha, Jiban Singh, Dorian Tosi Robinson, and Michael Vogel. The data and findings reported here are described in greater detail in the MSc theses of Dan Daniel, Moa Kenea, and Irfan Pratama (all of the IHE Delft Institute for Water Education).

Funding This research was funded by the Swiss Agency for Development Cooperation, the Eawag Partnership Program (EPP), and the REACH program funded by the UK Aid from the UK Department for International Development (DFID) for the benefit of developing countries (Aries Code 201880). The views expressed and information contained in this chapter are not necessarily those of or endorsed by these agencies, which can accept no responsibility for such views or information or for any reliance placed upon them.

References

1. WHO/UNICEF Joint Monitoring Programme (2015) Progress on drinking water, sanitation and hygiene: update and MDG assessment. WHO Press, Geneva/New York
2. Prüss-Ustün A, Bartram J, Clasen T, Colford JM, Cumming O, Curtis V, Bonjour S, Dangour AD, De France J, Fewtrell L et al (2014) Burden of disease from inadequate water, sanitation and hygiene in low and middle-income settings: a retrospective analysis of data from 145 countries. Trop Med Int Health 19:894–905
3. Montgomery MA, Bartram J, Elimelech M (2009) Increasing functional sustainability of water and sanitation supplies in Rural Sub-Saharan Africa. Environ Eng Sci 26(5):1017–1023
4. Marks SJ, Kumpel E, Guo J, Bartram J, Davis J (2018) Pathways to sustainability: a fuzzy-set qualitative comparative analysis of rural water supply programs. J Clean Prod 205:789–798
5. WHO/UNICEF Joint Monitoring Programme (2017) Progress on drinking water, sanitation and hygiene: 2017 update and SDG baselines. WHO Press, Geneva/New York
6. Daniel D (2019) Understanding the effect of socio-economic characteristics and psychosocial factors on household water treatment practices in rural Nepal using Bayesian Belief Networks. Int J Hyg Environ Health, in press.
7. Diener A, Schertenleib A, Daniel D, Kenea M, Pratama I, Bhatta M, Bhatta M, Marks SJ (2017) Adaptable drinking-water laboratory unit for decentralised testing in remote and alpine regions. In: 40th WEDC international conference, Loughborough, UK, pp 1–6
8. Tosi Robinson D, Schertenleib S, Kunwar B, Shrestha R, Bhatta M, Marks SJ (2018) Assessing the impact of a risk-based intervention on piped water quality in rural communities: the case of Mid-Western Nepal. Int J Environ Res Public Health 15:1616
9. United Nations Development Programme (2016) Human development report. New York, USA
10. Merz J, Nakarmi G, Shrestha SK, Dahal BM, Dangol PM, Dhakal MP, Dongol BS, Sharma S, Shah PB, Weingartner R (2003) Water: a scarce resource in rural watersheds of Nepal's middle mountains. Mt Res Dev 23:41–49
11. Udmale P, Ishidaira H, Thapa B, Shakya N (2016) The status of domestic water demand: supply deficit in the Kathmandu Valley, Nepal. Water 8:196
12. Xu J, Grumbine RE, Shrestha A, Eriksson M, Yang X, Wang YUN, Wilkes A (2009) The melting Himalayas: cascading effects of climate change on water, biodiversity, and livelihoods. Conserv Biol 23:520–530
13. Government of Nepal (2012) Poverty in Nepal 2010/11. Central Bureau of Statistics, Kathmandu
14. Helvetas Swiss Intercooperation (2014) Water: our work, Zurich, Switzerland
15. World Health Organization (2011) Guidelines for drinking water quality, 4th edn. WHO Press, Geneva
16. Daniel D, Diener A, van de Vossenberg J, Bhatta M, Marks SJ. Assessing drinking water quality at the point of collection and within household storage containers in the hilly rural area of Mid-Western Nepal, in review.

Dr. Sara Marks is an environmental engineer and research scientist at the Swiss Federal Institute of Aquatic Science and Technology (Eawag), Department of Sanitation, Water and Solid Waste for Development (Sandec) in Dübendorf, Switzerland. She has over a decade of experience conducting applied research on water, sanitation, and health, and she has collaborated with academics and practitioners in 12 countries worldwide. Her research group at Eawag focuses on evaluating and advancing technologies and systems for the effective treatment of drinking water, especially in low- and middle-income countries. Dr. Marks' projects have included implementation and assessment of water safety plans in Uganda and Nepal, evaluations of programs for the delivery of multiple-use water services in Burkina Faso and Tanzania, and development of promising technologies for water quality testing in challenging field conditions. Dr. Marks is thrilled to contribute to this book, since such a resource was not readily available to her as she navigated her own career path.

She found her calling in the field of environmental engineering,which allows her to harness inter-disciplinary methods to tackle issues at the intersection of water quality, technology, and public health. She believes that creativity and determination, combined with the tools of science and engineering, can help to tackle some of the most pressing challenges we face as a global community today.

Rubika Shrestha is a civil engineer with extensive practical experience working in the water, sanitation, and hygiene (WASH) sector in Nepal. She has a keen interest in applied research methods for the protection of public health, especially through improving WASH access and mitigating climate change impacts on freshwater resources. As the planning, monitoring, and knowledge management coordinator of the Helvetas Nepal Water Resources Management Programme (WARM-P), her key responsibilities are preparing, planning, and monitoring program activities, tracking progress of program activities, database management, report preparation, and knowledge dissemination. As the project coordinator of the research described in this chapter, Ms. Shrestha manages field activities, supports community training activities, develops survey instruments, and oversees an integrated data collection and management system. Previously, she worked as a WASH advisor in rural water supply and sanitation projects in western Nepal through a bilateral project of the government of Finland and the government of Nepal. In this capacity, she managed the project within Parbat districts. She also worked as an engineer in the District Development Committee, Tanahun, to implement local WASH projects. She has expertise in rural drinking water supply implementation, context-specific water safety plans, water quality testing tools, capacity building approaches, database management, and reporting. She pursued her bachelor's degree in civil engineering from Pulchowk Engineering Campus in Kathmandu, Nepal, and MSc in water resources management from Leuphana University, Suderburg Campus in Germany. She has been affiliated as a Professional Engineer by the Nepal Engineering Association (NEA).

Chapter 4
Water Quality for Decentralized Use of Non-potable Water Sources

Sybil Sharvelle

Abstract Pressures on water resources continue to rise with increasing population and diminishing local freshwater supplies. Use of locally available water sources can increase reliability and resilience of water supplies, particularly in areas prone to drought. Examples of local water supplies include roof runoff, stormwater, graywater, and treated wastewater. Conventionally, municipalities withdraw water from freshwater sources, treat that water to potable quality to meet urban water demand, and then treat and discharge wastewater. This approach results in substantial use of energy and consumables. Water is essentially imported and exported from local areas. An alternative is to use locally available water sources. This practice is gaining interest as an approach to minimize the import and export of water ensure reliable water sources, increase water supply resiliency, and promote energy efficiency. Local water sources are often supplied via decentralized water systems.

4.1 Introduction

Pressures on water resources continue to rise with increasing population and diminishing local freshwater supplies. Use of locally available water sources can increase reliability and resilience of water supplies, particularly in areas prone to drought. Examples of local water supplies include roof runoff, stormwater, graywater, and treated wastewater. Conventionally, municipalities withdraw water from freshwater sources, treat that water to potable quality to meet urban water demand, and then treat and discharge wastewater. This approach results in substantial use of energy and consumables. Water is essentially imported and exported from local areas. An alternative is to use locally available water sources. This practice is gaining interest as an approach to minimize the import and export of water [14], ensure reliable water sources, increase water supply resiliency, and promote energy efficiency. Local water sources are often supplied via decentralized water systems.

S. Sharvelle (✉)
Department of Civil and Environmental Engineering, Colorado State University,
Fort Collins, CO, USA
e-mail: Sybil.Sharvelle@Colostate.edu

© Springer Nature Switzerland AG 2020
D. J. O'Bannon (ed.), *Women in Water Quality*, Women in Engineering
and Science, https://doi.org/10.1007/978-3-030-17819-2_4

4.2   Decentralized Water Systems

Decentralized systems that make use of local water sources can use multiple source waters for varying end uses and can be applied from the building to neighborhood scale (Fig. 4.1). Common water sources for decentralized water systems and their definitions are included in Table 4.1. Decentralized non-potable water systems have been defined as systems in which local sources of water (e.g., roof runoff, stormwater, graywater, and wastewater) are collected, treated, and used for non-potable applications at the building, neighborhood, and/or district scale, generally at a location near the point of generation of the source of water [18].

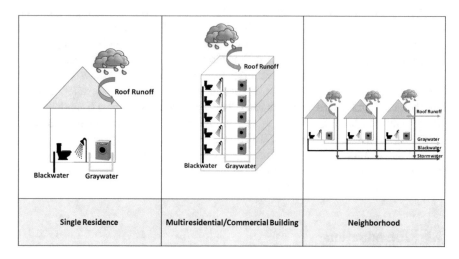

Fig. 4.1 Scales of decentralized water systems

Table 4.1 Water sources for decentralized water systems as defined by Sharvelle et al. [18]

Water source	Definition
Blackwater	Wastewater originating from toilets and/or kitchen sources (i.e., kitchen sinks and dishwashers)
Graywater	Wastewater collected from non-blackwater sources, such as bathroom sinks, showers, bathtubs, clothes washers, and laundry sinks
Wastewater	Water that is collected from combined graywater and blackwater sources, also known as sewage
Roof runoff	Precipitation from rain or snowmelt events collected directly off a roof surface that is not subject to frequent public access
Stormwater	Precipitation runoff from rain or snowmelt events that flows over land and/ or impervious surfaces (e.g., streets, parking lots, and rooftops)
Condensate	Water vapor that is converted to a liquid and collected, the most common source in buildings being air conditioning, refrigeration, and steam heating
Shallow groundwater	Groundwater located near the ground surface in an unconfined aquifer and, therefore, subject to contamination from infiltration of surface sources
Foundation water	Shallow groundwater collected from drainage around building foundations or sumps

Decentralized water (DW) systems can be applied at multiple local scales (Fig. 4.1). At the single residence scale, the most commonly used water sources are collected roof runoff and graywater. These water sources are typically used to meet irrigation demand or sometimes toilet flush demand at the single residence scale [14]. At the multi-residential and commercial scales, systems can be more complex when facilities management staff are on-site for management of the systems. At this scale, on-site-generated blackwater and wastewater can be used in addition to graywater and roof runoff, and use for toilet flushing and irrigation is most common (e.g., see DW Systems below). District systems are also feasible where water can be collected in a cluster multiple mixed-use (residential and commercial) buildings and used to meet water demand in those buildings. Neighborhood water systems (Fig. 4.1) can be classified as DW systems and would likely be locally operated with extensive online monitoring and control and oversight from municipalities [18]. Stormwater is more easily collected at this scale than the other scales discussed because it is collected from multiple impervious sources (including streets and parking lots; see Table 4.1). In addition, treatment of stormwater for non-potable use is pragmatic via neighborhood scale/decentralized systems.

4.3   Example DW Systems

4.3.1   Solaire Building (New York, NY)

The Solaire Building is a 27-story apartment building located in Battery Park, New York, NY. A membrane bioreactor located in the basement of the building treats wastewater collected in the building to use it for toilet flushing, heating, ventilation and air-conditioning (HVAC), and subsurface irrigation [23]. The system has been successfully operating since 2004 (Fig. 4.2).

Fig. 4.2 Solaire building and on-site treatment

4.4 San Francisco Public Utilities Commission Headquarters (San Francisco, CA)

The San Francisco Public Utilities Commission headquarters building houses approximately 950 employees and includes systems to use both blackwater and roof runoff on-site [16]. A Living Machine®, a plant-based, biological treatment system, treats wastewater to be used to flush toilets (Fig. 4.3). The building also has a 25,000 gallon cistern to collect roof runoff and use that water for subsurface irrigation. The building has reduced its demand for potable water by 65% through use of wastewater and roof runoff with almost all of the reduction achieved via use of wastewater for toilet flushing (Fig. 4.3).

4.5 Incentives for Decentralized Water Systems

There are many benefits to DW systems that serve as drivers to move these projects forward, even when they increase costs for developers. Motivation for implementation can include potable water demand reduction, commitment to sustainability and green building initiatives, stormwater runoff reduction, flexible infrastructure, and system-level energy benefits. San Francisco Public Utilities Commission [16] summarized 14 DW projects, and the most common drivers were to achieve Leadership in Energy and Environmental Design (LEED) certification and to meet city ordinances for stormwater runoff intensity and reduction requirements. While there are examples of DW systems that achieve cost savings (e.g., Solaire building), many systems installed to date do incur costs on developers, residents, or occupants [16]. However, there are many benefits that result in willingness to pay.

Fig. 4.3 SFPUC building and outdoor constructed wetlands for onsite treatment

4.5.1 Potable Water Demand Reduction

The use of on-site water sources reduces the demand for municipally supplied water, thereby diversifying water sources and increasing reliable, local water supplies. Of the 14 DW projects implemented in San Francisco, CA, potable water use reduction up to 81% was observed. More than half of these projects measured potable water use savings, achieving more than 20% reduction in use [16].

A study analyzing use of graywater and roof runoff scenarios to meet residential water demand in six cities across the USA showed potential demand reduction between 13% and 26% when graywater is used for irrigation and toilet flushing and up to 28% when roof runoff is captured using 2200 gallons of storage tanks and used depending on the local climate [14].

4.5.2 Sustainability and Green Building Initiatives

Aggressive sustainability goals have resulted in the boom of green building initiatives. The US Green Building Council developed the LEED certification program to rate buildings based on environmental performance. LEED certification points can be achieved by offsetting demand for potable water by using alternate water sources or via projects that result in decreased stormwater runoff volume: use of roof runoff or stormwater to meet on-site water demand achieves both of these goals. Many cities have ordinances in place to require new construction to be LEED certified, and there is a growing demand for potential tenants and the public. Of the 14 documented DW projects in San Francisco, LEED certification was a driver for the design for 8 of the 14 projects.

4.5.3 Stormwater Runoff Reduction

An important goal of stormwater control is to reduce stormwater runoff intensity and volume, particularly in cities with combined sewers. DW projects that capture stormwater or roof runoff for beneficial use achieve runoff intensity and volume reduction. Of note is that projects that use water sources other than stormwater do not achieve this goal. Many cities have ordinances or incentives in place to promote projects that reduce stormwater runoff volume and intensity. San Francisco has such an ordinance and is the driver for 11 of the 14 existing DW projects [16]. The District of Columbia Department of Energy and Environment has a stormwater retention credit trading program that has resulted in many projects that use roof and stormwater runoff in Washington, D.C. [14].

4.5.4   Infrastructure

Municipal-scale reclaimed water programs are infrastructure-intensive. In contrast, decentralized use of local water supplies provides the opportunity to diversify water sources without extensive public infrastructure requirements [21]. Decentralized water recycling was found to have cost and energy advantages compared to centralized systems in high growth areas where existing transmission infrastructure is nearing capacity limits [21]. While centralized treatment systems benefit from economy of scale for cost and energy efficiency, extensive requirements exist for conveyance infrastructure including pipes and pumping. Thus, life cycle costs can be reduced via decentralized infrastructure, particularly when the number of persons served is optimized to balance conveyance infrastructure requirements with scale of treatment systems [1, 8]. Decentralized infrastructure provides flexibility in where and when infrastructure is installed, avoiding unnecessarily overdesigning infrastructure.

4.5.5   System-Level Energy Efficiency

Water conservation is coupled with a decrease in energy consumption [20]. However, the interactions between DW systems and energy consumption are complex and multidimensional [14]. There are both cases where DW systems result in increased energy footprint and those where the systems result in energy efficiency. DW systems suffer from economy of scale and diversion of organic matter from centralized systems where that organic matter may have been converted to energy at a centralized wastewater treatment facility via anaerobic digestion. Water supply and wastewater treatment benefit from energy efficiency at larger scales when considering energy per gallon treated [5]. Thus, DW systems can suffer from economies of scale when considering energy efficiency [19, 21]. DW systems can also decrease energy footprint when organic matter is treated on-site and diverted from wastewater collection systems. Removing blackwater from wastewater collection systems that include anaerobic digestion and energy production from biosolids has high impact on life cycle assessment energy considerations [19, 22]. Thus, a scenario that assessed on-site graywater treatment and use with blackwater energy recovery was more energy and carbon efficient compared to a conventional, centralized system [22].

DW systems can offer energy efficiency compared to centralized systems in some cases. Energy efficiency is achieved when water sources can be used on-site with minimal or no treatment and when heat is recovered from the collected water source. When passive DW treatment systems are used that do not require energy for pumping or treatment, energy efficiency can be achieved [14] resulting in system-level energy benefits from reduced demand for potable water. In addition, when wastewater or graywater are collected on-site, there is potential for recovery of heat from those water sources resulting in the ability to meet up to 85% of the residential energy demand for heating water [10, 11]. On-site heat recovery from collected wastewater and improvements in energy efficiency of DW treatment systems offer the potential to achieve system-level energy benefits.

4.6 Current State of Regulations and Water Quality Requirements for Alternate Water Sources

Regulatory frameworks that allow for DW systems are sparse [18]. The cities of Los Angeles and San Francisco are examples of cities with programs in place to promote use of on-site water sources to meet non-potable demands [2, 9]. There are states that allow use of stormwater, roof runoff, and graywater as decentralized alternatives to meet non-potable demands (Tables 4.2 and 4.3). Non-potable demands included in current regulations are unrestricted irrigation and indoor use (e.g., toilet flushing and laundry), which have different water quality requirements in most jurisdictions. Unrestricted irrigation is defined here as water use for landscape where there may be human contact with irrigation water (e.g., spray irrigation in areas unrestricted to public access by barriers). The most common non-potable indoor use of alternate water sources is toilet flushing, but these water sources can also be used for laundry.

There is a wide range of water quality limits for indicator organisms (i.e., *E. coli* and total coliforms; Tables 4.2 and 4.3). The regulatory community has lacked guidance to set appropriate water quality standards for varying source water end-use combinations. For example, *E. coli* water quality targets for use of stormwater for unrestricted irrigation range from 2.2–4615 CFU/100 mL and total coliforms for treated graywater use to flush toilets range from 2.2–500 CFU/100 mL. Consistent guidance is needed to inform fit-for-purpose water quality standards for the use of different water sources.

The approach for water quality standards is end-point analysis of water quality, with the exception of the City of San Francisco, which has a non-potable water program in place that uses the risk-based framework developed by Sharvelle et al. [18]. End-point analysis relies on periodic analysis of water quality parameters in

Table 4.2 Summary of jurisdictions that allow use of alternate water sources for unrestricted irrigation and ranges of water quality limits in those jurisdictions (graywater is not included due to lack of regulations to allow this use (more commonly used for subsurface irrigation))

	Roof runoff[a]	Stormwater[b]	Blackwater/wastewater[c]
Jurisdictions with regulations for decentralized use	Los Angeles, CA; San Francisco, CA	Minnesota; District of Columbia; Los Angeles, CA; San Francisco, CA	San Francisco, CA
BOD_5 (mg/L)	NR	10	10
Turbidity (NTU)	10	2–3	Varies depending on treatment
TSS (mg/L)	NR	5–10	10
E. coli (CFU/100 mL)	100	2.2–4615	NR
Total coliforms (CFU/100 mL)	2.2	NR	2.2

[a]City and County of San Francisco [2]; Los Angeles County Department of Public Health [9]; NRC [14]
[b]Minnesota Pollution Control Agency [12]; DDOE [4]; City and County of San Francisco [2]; Los Angeles County Department of Public Health [9]
[c]City and County of San Francisco [2]

Table 4.3 Summary of jurisdictions that allow use of alternate water sources for indoor use (toilet flushing and sometimes laundry) and ranges of water quality limits in those jurisdictions

	Roof runoff[a]	Stormwater[b]	Graywater[c]	Blackwater/wastewater[d]
Jurisdictions with regulations for decentralized use	Los Angeles, CA; San Francisco, CA; CA; TX; GA	District of Columbia; Los Angeles, CA; San Francisco, CA	CA, CO, NM, OR, GE, TX, MS, WI	San Francisco, CA
BOD$_5$ (mg/L)	NR	10	10–200	10
Turbidity (NTU)	10	2	2–10	Varies depending on treatment system
TSS (mg/L)	NR	10	5–30	10
E. coli (CFU/100 mL)	100	2.2–50,000	14–200	NR
Total coliforms (CFU/100 mL)	2.2–500	NR	2.2–500	2.2

[a]City and County of San Francisco [2]; Los Angeles County Department of Public Health [9]
[b]DDOE [4]; City and County of San Francisco [2]; Los Angeles County Department of Public Health [9]
[c]NRC [14]; CDPHE [3]
[d]City and County of San Francisco [2]

the effluent, which can be problematic with respect to pathogens. End-point analysis of indicator organisms does not account for the specific loading of each virus, bacterium, and protozoan present in the source water and the expected level of treatment for those pathogens. In addition, indicator organisms may not reflect treatment of the specific pathogens present in the source water [6]. End-point analysis of pathogen concentrations in treated water can be cumbersome and expensive, without providing useful data regarding treatment of the source water to achieve acceptable health risk [18].

4.7 A Risk-Based Framework for the Development of Public Heath Guidance for DW Systems

Lack of a consistent regulatory framework to ensure protection of public health has been a barrier to wide implementation, but there are successful DW projects in place (e.g., see DW Systems above). An expert panel was convened to develop the Risk-Based Framework for the Development of Public Health Guidance for Decentralized Non-Potable Water Systems and chaired by Dr. Sybil Sharvelle [18]. The panel was organized by the National Water Research Institute.

The panel worked to achieve consensus on a framework that includes risk-based guidance developed from estimated pathogen log$_{10}$ reduction targets (LRTs) for various source waters and end-use combinations. The framework includes risk-based guidance on LRTs for pathogens (virus, bacteria, and protozoa), guidance for design to achieve LRTs, management, monitoring, and reporting for decentralized non-potable water (DNW) systems.

Quantitative Microbial Risk Assessment (QMRA; Fig. 4.4) was used to inform recommendations on LRTs for enteric viruses, enteric bacteria, and parasitic protozoa in various source water end-use combinations. QMRA relies on first characterizing pathogens in source waters (Fig. 4.4). Pathogen densities and occurrence for roof runoff, stormwater, graywater, and domestic wastewater were characterized by Schoen et al. [17] and Jahne et al. [7]. Exposure to water is then estimated based on the end use [15], which includes an estimate of frequency of exposure and quantity of water ingested during exposure. Next, an acceptable level of risk is selected for human infection or illness (Fig. 4.4). LRTs for the risk-based framework for DNW systems were established based on tolerable infection risks of 10^{-4} (1 in 1000 persons per year) or 10^{-2} (1 in 100 persons per year) per person for enteric viruses, enteric bacteria, and parasitic protozoa. These targets correspond to US Environmental Protection Agency's target pathogen risk for drinking water standards and recreational water quality standards, respectively. The 10^{-2} infection risk per person is typically used for involuntary exposures, while the 10^{-4} infection risk is typically used for voluntary exposures. Voluntary exposures occur when water users are aware that non-potable water is used and intentionally select to use that water. Involuntary exposures occur when the population may be exposed to non-potable water unknowingly and/or without an option to not be exposed. Multiple dose-response models were then used to estimate LRTs for the source water end-use combinations [17]. Unit processes can be selected to achieve those LRTs as described in Sharvelle et al. [18] based on developed LRTs for each source water end-use combination. Various unit processes achieve different log reduction values depending on how they are operated. For example, membrane filtration can achieve 2–6 log reductions of bacteria depending on pore size, and UV achieves 1–4 log reduction of bacteria for doses between 10 and 60 mJ/cm^2 [18] (Fig. 4.4).

4.7.1 A New Monitoring Approach

Water and wastewater systems have traditionally been monitored using fecal indicator organisms (FIOs). The presence or concentration of FIOs in samples of water or wastewater is considered indicative of other waterborne pathogens. FIOs have been

Fig. 4.4 Summary of quantitative microbial risk assessment

useful because they are expected to be present in water contaminated with fecal waste. The use of FIOs for a DNW system, however, has limitations, including:

- FIOs may not be present in potential source water for a non-potable system.
- FIOs are not necessarily representative of all pathogen groups.
- Grab samples analyzed for FIOs cannot be monitored continuously with high frequency.
- Measurement of FIOs requires hours to days and thus does not provide information on real-time performance.

The framework for DNW systems recommends that continuous water quality monitoring of surrogate parameters is conducted rather than a conventional monitoring system using grab samples for FIOs. Surrogate parameters such as turbidity, residual chlorine, ultraviolet transmittance, and others can be used to verify that unit processes are operating as they were designed to.

The three forms of monitoring for DNW systems include validation testing, field verification, and continuous verification monitoring. Each of these is defined below:

Validation testing. A treatment technology process evaluation study conducted where the system is challenged via addition of target or surrogate pathogens over a defined range of operating conditions, usually conducted at a test facility or in situ.

Field verification. A study to confirm that the installed treatment system performs as designed. Biological and/or chemical surrogates are monitored during the field verification which is typically conducted during commissioning (operational period where water is not supplied to the end use) and may be repeated later. In some cases, indigenous organisms can be used for process verification.

Continuous monitoring. Ongoing verification of system performance using sensors for the continuous observation of selected parameters, including surrogate parameters correlated with pathogen LRT requirements.

Continuous log reduction value (LRV) verification is accomplished using monitoring data from inline sensors that detect a surrogate parameter, which correlates directly with a given process LRV, at a high sampling frequency. Surrogate parameters should be those that can be monitored reliably at a high frequency and that are correlated with performance for pathogen reduction. Possible surrogates include chlorine residual, color, electrical conductivity, turbidity, and oxidation-reduction potential. For example, increased turbidity from standard operation in the effluent from an ultrafilter could indicate that performance for pathogen reduction is compromised. More detail is provided in Sharvelle et al. [18] on selection of appropriate surrogate parameters. Data acquisition systems are used to collect and log process monitoring data in a local and/or online database at the frequency required to determine process compliance. Online process controls are needed to modify operations or switch to an alternative water source when unit process is detected to be out of compliance.

4.7.2 Application of the Risk-Based Framework for DNW Systems

The City of San Francisco has applied the risk-based framework for DNW systems in its regulatory framework for non-potable water systems [2]. In addition, a National Blue Ribbon Commission for Onsite Non-Potable Water Systems was formed and developed a Guidebook for Developing and Implementing Regulations for Onsite Non-potable Water Systems [13]. The risk-based framework developed by Sharvelle et al. [18] served as the basis for this widely adopted guidebook.

4.8 Summary

Regulatory and permitting barriers have impeded implementation of such projects in spite of growing public interest. Decentralized non-potable water systems that exist to date were enabled by champions for those projects in both the development and regulatory sectors. In many cases, a regulatory framework was nonexistent and those projects either were approved via variances or paved the way for a regulatory framework. QMRA has been used to develop guidance for LRTs for various source water end-use combinations at the decentralized scale. The QMRA approach overcomes weaknesses in the end-point analysis of fecal indicator organisms by providing targets for pathogen reduction based on human health outcomes. The framework could be adapted for application across multiple scales of projects from the building to municipal scale to enable fit-for-purpose treatment of water, i.e., treatment appropriate for the source water end-use combination applied.

Dr. Sharvelle has had the pleasure to know two amazing female mentors, her M.S. advisor Dr. JoAnn Silverstein and her Ph.D. advisor Dr. M. Katherine Banks. These women not only mentored her to become a better scientist and engineer but shaped her approach to both professional and personal pursuits. From Dr. Silverstein, Sybil learned that quiet perseverance never fails. And from Dr. Banks, Dr. Sharvelle learned lifelong skills from an amazing female leader who identified strategic initiatives and pursued those with a laser focus. Dr. Sharvelle is most thankful to have had the opportunity to have worked with these women.

References

1. Bradshaw JL, Luthy RG (2017) Modeling and optimization of recycled water systems to augment urban groundwater recharge through underutilized stormwater spreading basins. Environ Sci Technol 51:11809–11819
2. City and County of San Francisco (2017) Directors rules and regulations regarding the operation of alternate water source systems, https://www.sfdph.org/dph/files/EHSdocs/ehsWaterdocs/.../SFHC_12C_Rules.pdf

3. Colorado Department of Public Health and Environment (CDPHE) (2015) Graywater control regulation #86, 5 CCR 1002-86
4. District Department of the Environment (2013) Stormwater management guidebook, https://doee.dc.gov/swguidebook. p M-18
5. Electric Power Research Institute (EPRI) (2002) Water and sustainability (volume 4): US electricity consumption for water supply & treatment – the next half century, EPRI, Palo Alto, CA: 2000.1006787
6. Harwood VJ, Levine AD, Scott TM, Chivukula V, Lukasik J, Farrah SR, Rose JB (2005) Validity of the inidcator organism paradigm for pathogen reduction in reclaimed water and public health protection. Appl Environ Microbiol 71(6):3163–3170
7. Jahne M, Schoen M, Ashbolt N, Garland J (2017) Simulation of enteric pathogen concentrations in locally-collected graywater and wastewater for microbial risk assessments. Microb Risk Anal 5:44–52
8. Kavvada O, Nelson KL, Horvath A (2018) Spatial optimization for decentralized non-potable water reuse. Environ Res Lett 13:064001
9. Los Angeles County Department of Public Health (2016) Guidelines for alternate water sources: indoor and outdoor non-potable uses, https://publichealth.lacounty.gov/eh/docs/ep_cross_con_AltWaterSourcesGuideline.pdf
10. McNabola A, Shields K (2013) Efficient drain water heat recovery in horizontal domestic shower drains. Energy Build 59:44–49
11. Meggers F, Leibundgut H (2011) The potential of wastewater heat exergy: decentralized high-temperature recovery with a heat pump. Energy Build 43(2011):879–886
12. Minnesota Pollution Control Agency (2017) Minnesota Stormwater Manual, https://stormwater.pca.state.mn.us/index.php?title=Stormwater_re-use_and_rainwater_harvesting&redirect=no. Accessed 20 Feb 18
13. National Blue Ribbon Commission for Onsite Non-Potable Water Systems (2017) A guidebook for developing and implementing regulations for onsite non-potable water systems, http://www.uswateralliance.org/initiatives/commission/resources
14. National Research Council (NRC) (2016) Using graywater and stormwater to enhance local water supplies: an assessment of risks, costs, and benefits. National Research Council, National Academies Press, Washington, DC
15. NRMMC; EPHC; AHMC (2006) Australian guidelines for water recycling: managing health and environmental risks (Phase 1) – November 2006. Document No. 21. National Water Quality Management Strategy; Natural Resource Management Ministerial Council (NRMMC), Environment Protection (EPHC), Australian Health Ministers Conference (AHMC); Biotext Pty Ltd, Canberra, Australia
16. San Francisco's Public Utilities Commission (SFPUC) (2017) San Francisco's non-potable water system projects
17. Schoen M, Ashbolt NJ, Jahne M, Garland J (2017) Risk-based enteric pathogen reduction targets for non-potable and direct potable use of roof runoff, stormwater, graywater, and wastewater. Microb Risk Anal 5:32–43
18. Sharvelle S, Ashbolt N, Clerico E, Holquist R, Leverenz H, Olivieri A (2017) Risk based framework for the development of public health guidance for decentralized non-potable water systems, Water Environment and Reuse Foundation, 2017, SIWM10C15
19. Shehabi A, Stokes JR, Horvath A (2012) Energy and air emission implications of a decentralized wastewater system. Environ Res Lett 7(2012):024007
20. Spang ES, Holguin AJ, Loge FJ (2018) The estimated impact of California's urban water conservation mandate on electricity consumption and greenhouse gas emissions. Environ Res Lett 13:014016
21. Woods GJ, Kang D, Quntanar DR, Curley EF, Davis SE, Lansey KE, Arnold RG (2013) Centralized versus decentralized wastewater reclamation in the Houghton area of Tucson, Arizona. J Water Resour Plan Manag 139(3):313–324

22. Xue X, Hawkins TR, Schoen MW, Garland J, Ashbolt NJ (2016) Comparing the life cycle energy consumption, global warming and eutrophication potentials of several water and waste service options. Water 8:154
23. Zavodo M (2018) NYC high rise reuse proves decentralized system works, Water and Wastewater International, www.waterworld.com/articles/wwi/print/volume-21/issue-1/features/nyc-high-rise-reuse-proves-decentralized-system-works.html. Accessed 18 Feb 2018

Sybil Sharvelle considered her love for math and physics combined with her appreciation for natural water systems and interest in closed-loop ecological systems when selecting a major for her undergraduate studies. Those interests and passions led her to select a major in Environmental Engineering. The deeper Dr. Sharvelle got into the coursework for her major, the more she realized that it was a good fit. Dr. Sharvelle received a B.S. in Civil Engineering with emphasis on Environmental Engineering from University of Colorado in 1998. At that time, her interest in closed-loop systems and resource recovery and reuse had grown much stronger. She found the perfect project to pursue her M.S. degree at University of Colorado with Dr. JoAnn Silverstein as her advisor. This project bridged her interest in closed-loop systems with her lifelong interest in the space program. Dr. Sharvelle had the opportunity to develop a biological processor to treat graywater (i.e., sink wastewater, not toilet wastewater) for reuse on the International Space Station. This was her dream project, and she loved it.

Dr. Sharvelle had the opportunity to further expand her interests in closed-loop recycling systems upon completion of her M.S. degree. Dr. Sharvelle pursued a doctoral degree at Purdue University working with Dr. M. Katherine Banks as part of the NASA Specialized Center of Research and Training (NSCORT) in Advanced Life Support. The US$10 million center included 24 primary investigators from multiple disciplines to address all components of an advanced life support system (e.g., food, water, and air). She continued to focus on graywater recycling for potable (drinking) use during space missions but then had the opportunity to learn about and be engaged with all of the other research supporting the NSCORT. This was a tremendous education and professional development opportunity for which Dr. Sharvelle will be forever grateful.

Dr. Sharvelle began her career as an Assistant Professor at Colorado State University in 2007 and was promoted to Associate Professor in 2017. There, she had the pleasure of working with Dr. Larry Roesner who had established a successful research program in the area of graywater reuse and served as an important early career mentor. In collaboration with Dr. Roesner, Dr. Sharvelle was able to launch her academic career with funding from Water Environment Research Foundation and Water Reuse Foundation to study the impacts of graywater use for landscape irrigation, conduct research on graywater treatment systems for toilet flushing, and evaluate public health and regulatory issues associated with graywater reuse. Through her research on graywater reuse, it became apparent to her that barriers associated with use of graywater and other non-potable water sources were not technical, but rather related to regulatory and social acceptance. Dr. Sharvelle became very interested in fit-for-purpose water standards for use on non-potable water sources. This led to her serving as a panel member for the National Research Council Panel on Using Graywater and Stormwater to Enhance Local Water Supplies: An Assessment of Risks, Costs, and Benefits and subsequently chairing the National Water Research Institute Panel to develop a Risk-Based Framework for the Development of Public Health Guidance for Decentralized Non-Potable Water Systems. The work proposed a paradigm shift from end-point analysis of water quality focused on monitoring of indicator organisms to a health risk–based approach to design systems that achieve appropriate reductions of pathogens specific to the source water end use.

Chapter 5
Wastewater-Based Epidemiology for Early Detection of Viral Outbreaks

Irene Xagoraraki and Evan O'Brien

Abstract The immense global burden of infectious disease outbreaks and the need to establish prediction and prevention systems have been recognized by the World Health Organization (WHO), the National Institutes of Health (NIH), the United States Agency of International Development (USAID), the Bill and Melinda Gates Foundation, and the international scientific community. Despite multiple efforts, this infectious burden is still increasing. For example, it has been reported that between 1.5 and 12 million people die each year from waterborne diseases and diarrheal diseases are listed within the top 15 leading causes of death worldwide. Rapid population growth, climate change, natural disasters, immigration, globalization, and the corresponding sanitation and waste management challenges are expected to intensify the problem in the years to come.

5.1 Introduction

The immense global burden of infectious disease outbreaks and the need to establish prediction and prevention systems have been recognized by the World Health Organization (WHO), the National Institutes of Health (NIH), the United States Agency of International Development (USAID), the Bill and Melinda Gates Foundation, and the international scientific community. Despite multiple efforts, this infectious burden is still increasing. For example, it has been reported that between 1.5 and 12 million people die each year from waterborne diseases [1, 2] and diarrheal diseases are listed within the top 15 leading causes of death worldwide [3]. Rapid population growth, climate change, natural disasters, immigration, globalization, and the corresponding sanitation and waste management challenges are expected to intensify the problem in the years to come.

Most infectious disease outbreaks in the United States have been related to microbial agents [4–7]. In the vast majority of cases, the infectious agents have not

I. Xagoraraki (✉) · E. O'Brien
Michigan State University, East Lansing, MI, USA
e-mail: xagorara@egr.msu.edu

© Springer Nature Switzerland AG 2020
D. J. O'Bannon (ed.), *Women in Water Quality*, Women in Engineering and Science, https://doi.org/10.1007/978-3-030-17819-2_5

been identified. However, the Environmental Protection Agency (EPA) suggests that most outbreaks of unidentified etiology are caused by viruses [8]. Viruses have been cited as potentially the most important and hazardous pathogens found in wastewater [9] and are included in the EPA contaminant candidate list. Viruses can lead to serious health outcomes, especially for children, the elderly, and immunocompromised individuals, and are of great concern because of their low infectious dose, ability to mutate, inability to be treated by antibiotics, resistance to disinfection, small size that facilitates environmental transport, and high survivability in water and solids.

Infectious outbreaks can cause uncontrollable negative effects especially in dense urban areas. Traditional disease detection and management systems are based on diagnostic analyses of clinical samples. However, these systems fail to detect early warnings of public health threats at a wide population level and fail to predict outbreaks in a timely manner. Classic epidemiology observes disease outbreaks based on clinical symptoms and infection status but does not have the ability to predict "critical locations" and "critical moments" for viral disease onset. Recent research efforts in developing optimized detection systems focus on rapid methods for analyzing blood samples, but this approach assumes that patients are examined at a clinical setting after the outbreak has been established and recognized.

The central premise of the proposed approach is that community wastewater represents a snapshot of the status of public health. Wastewater analysis is equivalent to obtaining and analyzing a community-based urine and fecal sample. Monitoring temporal changes in virus concentration and diversity excreted in community wastewater, in combination with monitoring metabolites and biomarkers for population adjustments, allows early detection of outbreaks (critical moments for the onset of an outbreak). In addition, carefully designed spatial sampling will allow detection of locations where an outbreak may begin to develop and spread (critical locations for the onset of an outbreak) (Fig. 5.1).

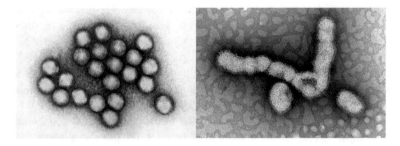

Fig. 5.1 Photomicrograph of adenovirus particles (left) and influenza virus particles (right). Adenovirus image from Dr. G. William Gary, Jr./CDC and influenza image from National Institute of Allergy and Infectious Diseases

5.2 Background

Similar detection systems have been used for the investigation of illicit drugs in various locations around the world [10–12]. The approach was first theorized in 2001 [10] and first implemented and reported for several illicit drugs in 2005 where the method was termed sewage epidemiology [11]. The methodology considers raw untreated wastewater as a reservoir of human excretion products; among these products are the parent compounds and metabolites of illicit drugs. If these excretion products are stable in wastewater as they travel through the sewage system, then the measured concentration from a wastewater treatment plant (WWTP) could correspond to the amount excreted by the serviced population. Table 5.1 presents a summary of prior studies utilizing the wastewater-based epidemiology methods to assess levels of various substances in a population.

Any substance that is excreted by humans and is stable (or has known kinetic pathways) in wastewater can be back-calculated into an initial source concentration. An important step in the application of wastewater-based epidemiology is the estimation of the contributing population and its sampled wastewater. Both census and biomarker data can be used in this approach to estimate the number of individuals that contribute to the wastewater sample.

5.3 Occurrence of Viruses in Wastewater

Waterborne viruses comprise a significant component of wastewater microbiota and are known to be responsible for disease outbreaks. A critical characteristic of viruses is that they do not grow outside the host cells. Therefore, viral concentrations in the wastewater stream will represent the concentrations excreted by the corresponding human population. Table 5.2 summarizes studies that detected waterborne and non-waterborne viruses in wastewater and human excrement.

Table 5.1 Summary of substances investigated via wastewater-based epidemiology

Substance	Country	References
Alcohol	Norway	[13]
Amphetamines	Australia, Belgium, Italy, Spain, South Korea, the United Kingdom, the United States	[14]
Cocaine	Australia, Belgium, Germany, Ireland, Italy, Spain, the United Kingdom, the United States	[14]
Counterfeit medicine	The Netherlands	[15]
Opiates	Germany, Italy, Spain, South Korea	[16]
Tobacco	Italy	[17]

Table 5.2 Summary of human viruses detected in wastewater or human excrement

Virus	Detected in excrement	Detected in wastewater	Reported concentrations in wastewater (copies/L)	References
Adenoviruses	Yes	Yes	$6.0*10^2$ to $1.7*10^8$	[21–36]
Astroviruses	Yes	Yes	$4.0*10^4$ to $4.1*10^7$	[24, 29, 43, 44, 47, 50–54]
Enteroviruses	Yes	Yes	$6.9*10^2$ to $4.7*10^6$	[24–26, 28, 29, 31, 33, 34, 36, 43, 53, 58–63]
Hepatitis A virus	Yes	Yes	$4.3*10^3$ to $8.9*10^5$	[29, 30, 58, 67–72]
Hepatitis E virus	Yes	Yes	$7.8*10^4$	[21, 34, 75–78]
Noroviruses	Yes	Yes	$4.9*10^3$ to $9.3*10^6$	[24–26, 28–30, 32, 33, 43, 47, 53, 54, 59, 60, 79, 84–88]
Rotaviruses	Yes	Yes	$1.8*10^3$ to $8.7*10^5$	[29–31, 36, 43–45, 47, 50, 53, 58, 59, 90–95]
Aichi virus	Yes	Yes	$9.7*10^4$ to $2.0*10^6$	[36, 96, 114]
Polyomaviruses	Yes	Yes	$8.3*10^1$ to $5.7*10^8$	[24, 35, 36, 85, 99, 115, 116]
Salivirus	Yes	Yes	$3.7*10^5$ to $9.7*10^6$	[97, 114, 117]
Sapovirus	Yes	Yes	$1.0*10^5$ to $5.1*10^5$	[24, 36, 98, 118]
Torque teno virus	Yes	Yes	$4.0*10^4$ to $5.0*10^5$	[23, 24, 35, 119]
Coronaviruses	Yes	Yes		[120, 121]
Influenza	Yes	Yes		[104–107, 122]
Dengue virus	Yes			[111–113, 123]
West Nile virus	Yes			[109, 110, 124]
Zika virus	Yes			[108, 125]
Yellow fever virus	Yes			[126, 127]

Note: The primary method of laboratory detection in the studies presented in this table is polymerase chain reaction (PCR), as well as real-time quantitative PCR (qPCR). PCR uses specific primers to replicate target sequences of nucleic acids; designing a primer to replicate a specific sequence in a given viral genome allows for the detection of that particular virus. qPCR can also determine the concentration of a virus in a sample by quantifying the number of copies of the target sequence

5.3.1 Waterborne Viruses

There are several groups of commonly detected and studied waterborne viruses, including adenoviruses, astroviruses, enteroviruses, hepatitis A and E viruses, noroviruses, and rotaviruses. Adenoviruses are known to cause gastroenteritis and respiratory disease [18] and have been linked to outbreaks of disease [19, 20].

Adenoviruses are a commonly studied group of viruses in water. They are commonly detected in raw wastewater [21–36] and have been cited as among the most significantly abundant human viruses in wastewater [24, 27, 28, 33, 37]. Adenoviruses have also been detected in human excrement of infected persons, including both feces and urine [38–47]. Studies have found the concentration of adenovirus in the stool of infected persons to range from 10^2 to 10^{11} copies per gram with an average concentration in the range from 10^5 to 10^6 copies per gram of stool [39, 41, 42, 46] as quantified by qPCR.

Astroviruses are a group of RNA viruses that have been linked to outbreaks of gastroenteritis [19, 48]. They have been cited as one of the more important viruses associated with gastroenteritis [49], but they have not been as commonly studied in wastewater compared to other groups of human enteric viruses. Nonetheless, they have been detected using standard PCR in wastewater in prior studies [24, 29, 50, 51]. They have also been detected in clinical samples of human excrement of infected people [43, 44, 47, 52–54], making them a viable candidate for wastewater epidemiology. While qPCR has been used as a detection method for astroviruses in human feces [44, 47, 54], and for quantification purposes in wastewater [51], no cited studies have reported quantitative values for astroviruses in human excrement.

Enteroviruses comprise several types of human enteric viruses, including polioviruses, coxsackieviruses, and echoviruses [55, 56]. Enteroviruses can cause an array of afflictions depending on type, including common cold, meningitis, and poliomyelitis [57], and have been linked to outbreaks of these diseases [19]. Enteroviruses have been detected via PCR in raw wastewater by numerous studies [25, 26, 28, 29, 31, 33, 34, 58, 59], as well as detected in human feces [43, 53, 60–63]. qPCR has not as yet been extensively employed to quantify enteroviruses in stool samples, though one study determined the enterovirus load to be in the range of $1.4*10^4$ to $6.6*10^9$ copies per gram of stool [60].

Two species of hepatitis viruses, hepatitis A virus and hepatitis E virus, are considered to be waterborne viruses. Hepatitis is a liver disease that can cause numerous afflictions, including fever, nausea, and jaundice [64]. Hepatitis A virus has been linked to disease outbreaks [65], and it has been suggested that even low levels of viral water pollution can produce infection [66]. Hepatitis A virus is often detected via PCR in raw wastewater [29, 30, 58, 67, 68] and several studies have also detected the virus in human stool samples [69–72]. Like enteroviruses, there has not been significant investigation into the quantification of hepatitis A virus in stool, though one study reported values in the range of $3.6*10^5$ to $5.6*10^9$ copies per gram of stool [70].

Hepatitis E virus, meanwhile, has only recently begun to become a pathogen of interest compared to other waterborne human viruses [73]. Like hepatitis A, hepatitis E virus can cause liver disease with many of the same symptoms; in fact, hepatitis E is not clinically distinguishable from other types of viral hepatitis infection [74]. While not investigated to the extent of other human enteric viruses, hepatitis E virus has been detected via PCR in raw wastewater [21, 34, 75]. There have also been studies that have detected hepatitis E virus in human stool samples [76–78]. One such study also used

RT-qPCR to quantify the concentration of hepatitis E virus in stool and reported values in the range of 10^1 to 10^6 copies per μL of stool [77].

Noroviruses, also known as Norwalk-like viruses, are a genus of viruses within the *Caliciviridae* family. They are one of the more significant gastroenteritis-causing viral agents, considered to be a leading cause of the disease [79–81], and are commonly associated with disease outbreaks [19, 82, 83]. Noroviruses are one of the more commonly investigated and detected viruses in wastewater [24–26, 28–30, 32, 33, 36, 59, 60, 84, 85]. A number of studies have also investigated the presence of noroviruses in human feces [43, 47, 53, 54, 79, 86–88]. One such cited study reported quantification values for norovirus in stool following qPCR, in the range of $9.7*10^5$ to $1.1*10^{12}$ copies per gram, with a mean value of approximately 10^{11} copies per gram [87].

Rotaviruses are another primary cause of gastroenteritis with symptoms including diarrhea, vomiting, and fever, in accordance with other enteric viruses [89]. They are commonly detected via PCR in raw wastewater [29–31, 36, 50, 58, 59, 90–92] and are commonly investigated and detected in human feces [43–45, 47, 53, 93–95]. Like other waterborne viruses, though, only a handful of studies on rotaviruses have used qPCR as a detection tool, and none reported quantification values in terms of the number of copies.

In addition to the commonly investigated waterborne viruses described above, there are other human viruses that are commonly detected in wastewater and human stool but not as frequently studied, such as Aichi virus, polyomaviruses, salivirus, sapovirus, and torque teno virus. Aichi virus is a member of the *Picornaviridae* family, the same family as enteroviruses, and is believed to cause gastroenteritis [96]. Salivirus, another member of the *Picornaviridae* family, is also associated with gastroenteritis, as well as acute flaccid paralysis [97]. Sapovirus, like norovirus, is a member of the *Caliciviridae* family and like its relative is a common cause of gastroenteritis [98]. Polyomaviruses are associated with a variety of diseases in humans, including nephropathy, progressive multifocal leukoencephalopathy, and Mercel cell carcinoma [99]. Torque teno virus is commonly detected in humans, but the clinical consequences of infection are unclear [100]. These viruses are included in Table 5.2.

5.3.2   Non-waterborne Viruses

Non-waterborne viruses have also been detected in wastewater or human excrement (included in Table 5.2). While it is logical to investigate the applicability of waterborne viruses to wastewater-based epidemiology, it is also important to note the potential for other categories of viruses to fit into this methodology.

There exists a category of water-related viruses that are transmitted via insects (like mosquitos) that breed in water, such as Zika virus, West Nile virus, Rift Valley fever virus, yellow fever virus, dengue virus, and chikungunya virus, in addition to confirmed waterborne viruses. These viruses also fall into the category of zoonotic viruses, which are viruses that can be transmitted between humans and animals. Other zoonotic viruses include avian influenza virus, SARS (Severe Acute Respiratory Syndrome) coronavirus, Menangle virus, Tioman virus, Hendra virus, Australian bat lyssavirus, Nipah virus, and hantavirus. Specific animal species of concern that are vectors for these zoonotic viruses include avian species, bats, rodents, and mosquitos. While these zoonotic viruses are not classified as water-borne, they are associated with potential waterborne transmission, such as exposure to aerosolized wastewater, which can occur when wastewater undergoes turbulence, such as in flush toilets, converging sewer pipes, and aeration basins [101, 102] as well as irrigation and land application systems.

It has been shown that coronaviruses have been detected in wastewater [103] and SARS coronaviruses have been detected in stool and urine samples. Furthermore, detection in both human stool and urine [104–106] as well as wastewater [107] has been reported for influenza. Detection in urine has been reported for the mosquito-associated Zika virus [108], West Nile virus [109, 110], dengue virus [111, 112], and yellow fever virus [113]. These observations indicate that the concept of wastewater-based epidemiology could be applied to a wide range of viruses beyond the confirmed waterborne viruses.

5.4 Variations of Viruses in Wastewater

The quantity of human enteric viruses in wastewater has been shown to have seasonal variation, indicating that infection resulting from these viruses is more prevalent at certain times of the year. A study conducted in Japan by Katayama et al. (2008) found that norovirus concentrations in wastewater were highest during the months of November through April [26], while enterovirus and adenovirus concentrations were largely consistent throughout the year. A 9-year study in Milwaukee, Wisconsin, by Sedmak et al. (2005) found that concentrations of reoviruses, enteroviruses, and adenoviruses were highest during the months of July through December. This study also analyzed clinical specimens of enterovirus isolates and found the incidence of clinical enterovirus infection corresponded to the concentration of these viruses in wastewater during the same time periods [31]. Another study in Beijing, China, by Li et al. (2011) found that rotavirus concentrations were highest during the months of November through March [90] and that these findings also corresponded with clinical rotavirus data reported in China [128].

Additionally, variation in viral concentration in wastewater can occur on a smaller timescale. For example, tourist locations could experience higher wastewater loads, and consequently higher viral concentrations, on weekends where there is an influx of population. For example, Xagoraraki's research group conducted a study which observed an increase in adenovirus concentration in wastewater following the July 4th holiday in Traverse City, Michigan, a popular vacation destination [27]. Likewise, urban centers may experience higher loads during the day on weekdays, while people are at work. Accounting for these population changes would be vital for understanding when viral outbreaks occur.

Wastewater has been used in the past as a tool to investigate viruses for other purposes as well, such as spatial surveillance and evaluation of immunization efficacy. Two particular studies were able to use wastewater to observe the spatial variation of particular viral strains; Bofill-Mas et al. observed that particular strains of polyomavirus were endemic to specific regions, while Clemente-Cesares et al. detected Hepatitis E virus in areas previously considered non-endemic for the virus [129, 130]. Lago et al. (2003) investigated the efficacy of a poliovirus (a type of enterovirus) immunization campaign in Havana, Cuba, by quantifying concentrations of the virus in wastewater [61]. Poliovirus was detected in 100% of wastewater samples prior to the start of the immunization campaign and dropped to a 0% detection rate in wastewater 15 weeks after the campaign, indicating the usefulness of wastewater surveillance. A study by Carducci et al. (2006) investigated the relationship between wastewater samples and clinical samples and found that the same viral strains could sometimes be detected between the two sets of samples [131].

5.5   Proposed Methodology

Waterborne and non-waterborne viruses have been detected in wastewater, variations of concentrations in time have been observed, and virus presence in wastewater has on occasion been correlated with occurrence of clinical disease. However, wastewater-based epidemiology methods have not yet been applied to assess and predict viral disease outbreaks in a systematic way. Wastewater-based epidemiology has the potential to predict "critical locations" and "critical moments" for viral disease onset. Designing spatial and temporal sampling appropriate to the area of concern, as well as modeling the fate of viruses, is critical for the effectiveness of the proposed method. This methodology is summarized in Fig. 5.2. In the following sections, critical factors for implementation are discussed.

5.5.1   Sampling in Urban and Rural Locations

The most critical parameter for the effective application of wastewater-based epidemiology is the selection of a surveillance program, including spatial and temporal sampling. Considerations must be made in the differences between urban and rural wastewater systems. Urban sewage systems offer a convenient confluence of wastewater in the serviced population, as all wastewater will ultimately flow to a WWTP, providing a sampling point representing the entire community. Additionally, localized sampling can be performed in specific neighborhoods where access points are available. By surveying both the combined wastewater at the treatment plant and the localized samples from neighborhoods, viral outbreaks can be traced to a more

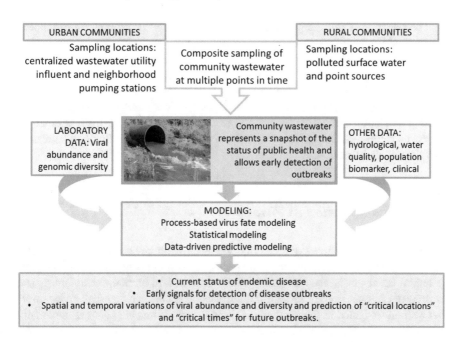

Fig. 5.2   Summary of the proposed wastewater-based epidemiology methodology

specific location and the urban areas of concern can be identified. Xagoraraki's research group is currently conducting an National Science Foundation-funded study of this nature in the city of Detroit, sampling at several interceptors at the Detroit wastewater treatment plant, as well as sampling from sewer lines in residential areas throughout the city.

More rural or underdeveloped areas that do not have sewage collection systems pose sampling problems. In these areas, wastewater is often disposed in open space, latrines, or septic tanks. As a result, for wastewater-based epidemiology sampling to be effectively applied to these areas, disposal, fate, and transport of wastewater in the environment must be taken into account. Watershed modeling would therefore become an integral component of the wastewater-based epidemiology methodology for rural locations. In a study performed by Xagoraraki's research group, preliminary investigation into the wastewater epidemiology methodology was conducted [132]. Samples were collected from a wastewater treatment plant and surrounding surface waters in Kampala, Uganda. Three sampling events were conducted in 2-week intervals. Four human viruses (adenovirus, enterovirus, hepatitis A virus, and rotavirus) were quantified at each sampling location via qPCR. Concentrations of each virus at each location from each sampling event were compared to one another to determine if significant differences could be observed from one sampling event to the next. Results indicated that statistically significant differences in viral concentration were observed for the measured viruses at several sampling locations.

The selection of the sampling times and locations is of paramount importance to the methodology, regardless of whether sampling takes place in urban or rural areas. Sampling should be based upon expected critical pathways of viral transport and transmission. These critical pathways include environmental reservoirs for viruses and the timing and locations where viruses are most easily transported and transmitted between humans and the environment. By determining sampling times and locations based upon critical pathways, "critical locations," and "critical moments," areas and times most impactful to the spread of viral disease would be most readily and effectively identified.

5.5.2 Quantification of Viruses

Quantitative data of viruses of concern, such as those obtained with qPCR, are critical for the proposed methodology, as peaks in viral concentrations will indicate potential onset of disease outbreaks. While detection in human excrement or raw wastewater has not been reported for all viruses, it is possible that they have simply not been investigated in this context, as detection of viruses via conventional methods (cell culture, PCR, qPCR) is specific to the virus being investigated. Thus, while qPCR is important to detect and quantify common waterborne viruses, next-generation sequencing and metagenomic methods could also be performed to screen for the presence of other viruses. If genomic sequences of viruses of concern are found, then quantification with qPCR can follow.

Metagenomic methods have been applied to investigate viruses in wastewater and have been found to produce more conservative results of viral detection compared to conventional methods; viruses detected with metagenomic methods are typically also detected with conventional methods, whereas viruses detected via qPCR may not be detected with metagenomic methods. These metagenomic methods, however, can detect the presence of viruses not commonly quantified using qPCR [37, 133–136]. Xagoraraki's research group's studies have used metagenomic methods to identify human viruses of potential concern in wastewater. The first of these studies, conducted with samples from both Michigan and France, detected a comparatively high number of metagenomic hits for human herpesviruses and also detected human parvovirus and human polyomavirus in wastewater effluents [37]. Their other study, conducted in Uganda, detected human astroviruses, papillomaviruses, as well as a BLAST (Basic Local Alignment Search Tool) hit for Ebola virus [132]. While more research is still required to attain more robust genomic information and comparison databases, metagenomic methods can still be a useful tool for the identification of potential viruses that can then be monitored with qPCR methods. Table 5.3 presents a summary of studies that have used metagenomic methods to detect human viruses in wastewater and human excrement.

Table 5.3 Summary of studies using metagenomic methods to detect viral sequences in wastewater and human excrement

Detected in	Virus	References
Wastewater	Adenovirus, enterovirus, polyomavirus, papillomavirus	[135]
	Adenovirus, Aichi virus, coronavirus, herpesvirus, torque teno virus	[137]
	Adenovirus, Aichi virus, astrovirus, coronavirus, enterovirus, herpesvirus, papillomavirus, parechovirus, parvovirus, rotavirus, salivirus, sapovirus, torque teno virus	[133]
	Adenovirus, Aichi virus, astrovirus, norovirus, papillomavirus, parechovirus, polyomavirus, salivirus, sapovirus	[136]
	Adenovirus, herpesvirus, parvovirus, polyomavirus	[37]
	Adenovirus, astrovirus, Ebola virus, enterovirus, papillomavirus, rotavirus, torque teno virus	[132]
Human excrement	Adenovirus, astrovirus, enterovirus, norovirus, parvovirus, rotavirus, torque teno virus	[138]
	Adenovirus, Aichi virus, enterovirus, parechovirus, rotavirus	[139]

Note: The following sequences have been confirmed via PCR for the listed study. Bibby and Peccia [133] adenovirus, enterovirus, parechovirus [131], [136], adenovirus, polyomavirus, salivirus [134], adenovirus [37], [132], adenovirus, enterovirus, rotavirus [130]

5.5.3 *Population Normalization*

Population normalization is also a critical factor for the application of wastewater-based epidemiology. Proper quantification of biomarkers in wastewater would allow for an appropriate estimation of serviced population via statistical modeling, which would provide context to measured viral concentrations and ensure that differences in viral concentration could not be attributed to changes in population. When observed viral concentrations are significantly high relative to the estimated population, a viral outbreak could be indicated.

Quantification of biomarkers (substances naturally excreted by humans) in wastewater can be used as a method of estimating population in an area. Governmental census information has been found to underestimate the population of a community compared to estimation using biomarkers [140], and certain substances detected in wastewater have been shown to correlate with census data [141]. Several substances have been proposed and investigated as population biomarkers (Table 5.4), including creatinine [142], cholesterol, coprostanol [143], nicotine [144], cortisol, androstenedione, and the serotonin metabolite 5-hydroxyindoleacetic acid (5-HIAA) [145]. Nutrients such as nitrogen, phosphorus, and oxygen [12], as well as ammonium [146], have also been proposed as population biomarkers, but these may more adequately reflect human activity and industry footprint rather than population [145, 147, 148].

Table 5.4 Summary of biomarkers proposed for population adjustment

Biomarker	Description	Excreted in	References
5-HIAA	Metabolite of serotonin	Urine	[145]
Ammonium	Form of ammonia found in water	Urine	[146]
Androstenedione	Sex hormone precursor	Urine	[149]
Atenolol	Drug (beta blocker) used to treat hypertension	Urine	[140]
Cholesterol	Lipid molecule, key component of cell membranes	Feces	[143]
Coprostanol	Metabolite of cholesterol	Feces	[143]
Cortisol	Steroid hormone produced by adrenal glands	Urine	[150]
Cotinine	Metabolite of nicotine	Urine	[145]
Creatinine	Metabolite of creatine phosphate in muscle	Urine	[142]
Nicotine	Stimulant found in tobacco	Urine	[144]
Nutrients (N, P, BOD)	Water-quality parameters	n/a	[12]

Table 5.5 Summary of reported shedding rates for viruses

Virus	Range of shedding rate, copies/g stool	References
Adenoviruses	$1.0*10^2$ to $1.0*10^{11}$	[39, 41, 42, 46]
Enteroviruses	$1.4*10^4$ to $6.6*10^9$	[60]
Hepatitis A virus	$3.6*10^5$ to $1.0*10^{11}$	[70]
Hepatitis E virus	$1.0*10^1$ to $1.0*10^6$	[77]
Noroviruses	$1.1*10^5$ to $1.1*10^{12}$	[87]
Sapoviruses	$1.3*10^5$ to $2.5*10^{11}$	[98, 118]

5.5.4 Estimation of Shedding Rates

The shedding rate (the rate with which viruses are released from the body in excrement) for each waterborne virus group encompasses a wide range, from 10^2 copies per gram at minimum to 10^{12} copies per gram at maximum. This variability is summarized for selected viruses in Table 5.5. For example, mean concentration values of adenoviruses in excrement ranged from 10^4 to 10^6 depending on the study and whether the virus is excreted in stool or urine, indicating a wide data variance [39, 41]. Many factors can impact the shedding rate of viruses in excrement, including viremia (the presence of the virus in the bloodstream) [40, 87, 151]. The duration of the presentation of a particular disease can also impact the shedding rate [105, 121].

5.5.5 Transport of Viruses in the Environment

Waterborne viruses survive well in water, but all viruses are susceptible to natural degradation determined by factors such as temperature, exposure to UV light, and the microbial community [152, 153]. The kinetic decay rate of a virus would thereby be primarily dependent not only on the characteristics of the individual virus but also environmental conditions within the sewage system, which could vary from location to location. Moreover, the fate of viruses may be different between wastewater systems in urban areas which typically use enclosed underground sewer pipes and rural areas which may utilize septic tanks, catchments, and the open environment. Viruses can also adsorb to or be enveloped by particulate matter in wastewater which would lead to confounding factors in measurement of these viruses.

5.5.6   Correlation with Public Health Records and Unidentified Clinical Data

Comparison with clinical data is another key component of these methods. Correlations between measured viral concentrations in wastewater and reported clinical cases of disease could be established, strengthening the proposed methodology. The establishment of these correlations can serve as a validation for a prediction model that accounts for the factors discussed above, providing evidence for the notion that changes of viral concentrations in wastewater will indicate changes in viral disease cases in humans. Moreover, should preventative public health measures be implemented after the identification of an outbreak, the tracking of clinical data could provide a quantifiable indicator of the efficacy of these preventative measures.

5.6   Conclusions

Infectious viral outbreaks can cause uncontrollable negative effects especially in densely populated areas. Early detection is critical for effective management and prevention of outbreaks. Recent research efforts in developing optimized detection systems often focus on rapid methods for analyzing blood or excrement samples; however, these approaches require that individuals are examined in clinical settings, typically after an outbreak has been established. Wastewater-based epidemiology is a promising methodology for early detection of viral outbreaks at a population level. Analyzing wastewater is equivalent to obtaining and analyzing a community excrement sample. In the determination of whether an outbreak is imminent or already in progress, quantifying viral concentration in raw wastewater is a crucial first step in this process. Waterborne viruses appear to be prime candidates, as they are detectable and quantifiable in both wastewater and human excrement. Non-waterborne viruses have been shown to be detected in human excrement, and some have been reported to be detected in wastewater. Wastewater-based epidemiology therefore has the potential to expand beyond waterborne viruses.

Routine monitoring for temporal changes in virus concentration and diversity in community wastewater, in combination with monitoring metabolites and biomarkers for population adjustments, allows early detection of outbreaks (critical moments for the onset of an outbreak). In addition, carefully designed spatial sampling of wastewater will allow detection of locations where an outbreak may begin to develop and spread (critical locations for the onset of an outbreak). Considerations in sampling locations must be taken with regard to the area of investigation, as urban and rural areas may have differences in the respective wastewater systems that can affect viral transport in the water environment. Moreover, to obtain an accurate estimation of disease cases in a population, other factors must be considered such as viral shedding rates, environmental transport and degradation rates, and correlation with reported clinical disease data. Ultimately, there is great opportunity for the use of wastewater-based epidemiology to investigate viral outbreaks within a community. Comprehensive application of the various factors discussed above is crucial for the full potential of this methodology to be realized. Further research could clarify many of these issues and allow for the full development and application of this new epidemiological technique for studying, identifying, and predicting viral outbreaks.

References

1. Gleick PH (2002) Dirty-water: estimated deaths from water-related diseases 2000–2020. Pacific Institute for Studies in Development, Environment, and Security, Oakland, CA. Citeseer
2. Prüss-Üstün A, Bos R, Gore F, Bartram J, World Health Organization. 2008. Safer water, better health: costs, benefits and sustainability of interventions to protect and promote health. World Health Organization, Geneva.
3. Mathers CD, Loncar D (2006) Projections of global mortality and burden of disease from 2002 to 2030. PLoS Med 3:e442
4. Barwick RS, Levy DA, Craun GF, Beach MJ, Calderon RL (2000) Surveillance for waterborne-disease outbreaks—United States, 1997–1998. MMWR CDC Surveill Summ 49:1–21
5. Kramer MH, Herwaldt BL, Craun GF, Calderon RL, Juranek DD (1996) Surveillance for waterborne-disease outbreaks–United States, 1993–1994. MMWR CDC Surveill Summ 45:1–33
6. Levy DA, Bens MS, Craun GF, Calderon RL, Herwaldt BL (1998) Surveillance for waterborne-disease outbreaks—United States, 1995–1996. MMWR CDC Surveill Summ: 47:1–34
7. Liang JL, Dziuban EJ, Craun GF, Hill V, Moore MR, Gelting RJ, Calderon RL, Beach MJ, Roy SL (2006) Surveillance for waterborne disease and outbreaks associated with drinking water and water not intended for drinking – United States, 2003–2004. Morb Mortal Wkly Rep 55:31–65
8. US EPA (2006) National primary drinking water regulations: ground water rule; final rule. Fed Regist 71:65574–65660
9. Toze S (1997) Microbial pathogens in wastewater: literature review for urban water systems multi-divisional research program. CSIRO Land and Water, Floreat Park
10. Daughton CG, Jones-Lepp TL (2001) Pharmaceuticals and personal care products in the environment: scientific and regulatory issues. American Chemical Society, Washington, DC
11. Zuccato E, Chiabrando C, Castiglioni S, Bagnati R, Fanelli R (2008) Estimating community drug abuse by wastewater analysis. Environ Health Perspect 116:1027–1032

12. van Nuijs ALN, Castiglioni S, Tarcomnicu I, Postigo C, de Alda ML, Neels H, Zuccato E, Barcelo D, Covaci A (2011) Illicit drug consumption estimations derived from wastewater analysis: a critical review. Sci Total Environ 409:3564–3577

13. Reid MJ, Langford KH, Mørland J, Thomas KV (2011) Analysis and interpretation of specific ethanol metabolites, ethyl sulfate, and ethyl glucuronide in sewage effluent for the quantitative measurement of regional alcohol consumption. Alcohol Clin Exp Res 35:1593–1599

14. Irvine RJ, Kostakis C, Felgate PD, Jaehne EJ, Chen C, White JM (2011) Population drug use in Australia: a wastewater analysis. Forensic Sci Int 210:69–73

15. Venhuis BJ, Voogt P, Emke E, Causanilles A, Keizers PHJ (2014) Success of rogue online pharmacies: sewage study of sildenafil in the Netherlands. BMJ 349:g4317

16. Hummel D, Löffler D, Fink G, Ternes TA (2006) Simultaneous determination of psychoactive drugs and their metabolites in aqueous matrices by liquid chromatography mass spectrometry. Environ Sci Technol 40:7321–7328

17. Castiglioni S, Senta I, Borsotti A, Davoli E, Zuccato E (2015) A novel approach for monitoring tobacco use in local communities by wastewater analysis. Tob Control 24:38–42

18. Ganesh A, Lin J (2013) Waterborne human pathogenic viruses of public health concern. Int J Environ Health Res 23:544–564

19. Maunula L, Klemola P, Kauppinen A, Söderberg K, Nguyen T, Pitkänen T, Kaijalainen S, Simonen ML, Miettinen IT, Lappalainen M, Laine J, Vuento R, Kuusi M, Roivainen M (2008) Enteric viruses in a large waterborne outbreak of acute gastroenteritis in finland. Food Environ Virol 1:31–36

20. Papapetropoulou M, Vantarakis AC (1998) Detection of adenovirus outbreak at a municipal swimming pool by nested PCR amplification. J Infect 36:101–103

21. Bofill-Mas S, Albinana-Gimenez N, Clemente-Casares P, Hundesa A, Rodriguez-Manzano J, Allard A, Calvo M, Girones R (2006) Quantification and stability of human adenoviruses and polyomavirus JCPyV in wastewater matrices. Appl Environ Microbiol 72:7894–7896

22. Bofill-Mas S, Rodriguez-Manzano J, Calgua B, Carratala A, Girones R (2010) Newly described human polyomaviruses Merkel Cell, KI and WU are present in urban sewage and may represent potential environmental contaminants. Virol J 7:141

23. Carducci A, Morici P, Pizzi F, Battistini R, Rovini E, Verani M (2008) Study of the viral removal efficiency in a urban wastewater treatment plant. Water Sci Technol 58:893–897

24. Hata A, Kitajima M, Katayama H (2013) Occurrence and reduction of human viruses, F-specific RNA coliphage genogroups and microbial indicators at a full-scale wastewater treatment plant in Japan. J Appl Microbiol 114:545–554

25. Hewitt J, Leonard M, Greening GE, Lewis GD (2011) Influence of wastewater treatment process and the population size on human virus profiles in wastewater. Water Res 45:6267–6276

26. Katayama H, Haramoto E, Oguma K, Yamashita H, Tajima A, Nakajima H, Ohgaki S (2008) One-year monthly quantitative survey of noroviruses, enteroviruses, and adenoviruses in wastewater collected from six plants in Japan. Water Res 42:1441–1448

27. Kuo DH-W, Simmons FJ, Blair S, Hart E, Rose JB, Xagoraraki I (2010) Assessment of human adenovirus removal in a full-scale membrane bioreactor treating municipal wastewater. Water Res 44:1520–1530

28. La Rosa G, Pourshaban M, Iaconelli M, Muscillo M (2010) Quantitative real-time PCR of enteric viruses in influent and effluent samples from wastewater treatment plants in Italy. Ann Ist Super Sanità 46:266–273

29. Petrinca AR, Donia D, Pierangeli A, Gabrieli R, Degener AM, Bonanni E, Diaco L, Cecchini G, Anastasi P, Divizia M (2009) Presence and environmental circulation of enteric viruses in three different wastewater treatment plants. J Appl Microbiol 106:1608–1617

30. Prado T, Silva DM, Guilayn WC, Rose TL, Gaspar AMC, Miagostovich MP (2011) Quantification and molecular characterization of enteric viruses detected in effluents from two hospital wastewater treatment plants. Water Res 45:1287–1297

31. Sedmak G, Bina D, MacDonald J, Couillard L (2005) Nine-year study of the occurrence of culturable viruses in source water for two drinking water treatment plants and the influent and

effluent of a wastewater treatment plant in Milwaukee, Wisconsin (August 1994 through July 2003). Appl Environ Microbiol 71:1042–1050

32. Sima LC, Schaeffer J, Saux J-CL, Parnaudeau S, Elimelech M, Guyader FSL (2011) Calicivirus removal in a membrane bioreactor wastewater treatment plant. Appl Environ Microbiol 77:5170–5177

33. Simmons FJ, Kuo DH-W, Xagoraraki I (2011) Removal of human enteric viruses by a full-scale membrane bioreactor during municipal wastewater processing. Water Res 45:2739–2750

34. Masclaux FG, Hotz P, Friedli D, Savova-Bianchi D, Oppliger A (2013) High occurrence of hepatitis E virus in samples from wastewater treatment plants in Switzerland and comparison with other enteric viruses. Water Res 47:5101–5109

35. Hamza IA, Jurzik L, Überla K, Wilhelm M (2011) Evaluation of pepper mild mottle virus, human picobirnavirus and Torque teno virus as indicators of fecal contamination in river water. Water Res 45:1358–1368

36. Kitajima M, Iker BC, Pepper IL, Gerba CP (2014) Relative abundance and treatment reduction of viruses during wastewater treatment processes—Identification of potential viral indicators. Sci Total Environ 488–489:290–296

37. O'Brien E, Munir M, Marsh T, Heran M, Lesage G, Tarabara VV, Xagoraraki I (2017a) Diversity of DNA viruses in effluents of membrane bioreactors in Traverse City, MI (USA) and La Grande Motte (France). Water Res 111:338–345

38. Allard A, Albinsson B, Wadell G (1992) Detection of adenoviruses in stools from healthy persons and patients with diarrhea by two-step polymerase chain reaction. J Med Virol 37:149–157

39. Berciaud S, Rayne F, Kassab S, Jubert C, Faure-Della Corte M, Salin F, Wodrich H, Lafon ME (2012) Adenovirus infections in Bordeaux University Hospital 2008–2010: clinical and virological features. J Clin Virol 54:302–307

40. Heim A, Ebnet C, Harste G, Pring-Åkerblom P (2003) Rapid and quantitative detection of human adenovirus DNA by real-time PCR. J Med Virol 70:228–239

41. Jeulin H, Salmon A, Bordigoni P, Venard V (2011) Diagnostic value of quantitative PCR for adenovirus detection in stool samples as compared with antigen detection and cell culture in haematopoietic stem cell transplant recipients. Clin Microbiol Infect 17:1674–1680

42. Lion T, Kosulin K, Landlinger C, Rauch M, Preuner S, Jugovic D, Pötschger U, Lawitschka A, Peters C, Fritsch G, Matthes-Martin S (2010) Monitoring of adenovirus load in stool by real-time PCR permits early detection of impending invasive infection in patients after allogeneic stem cell transplantation. Leukemia 24:706–714

43. Martínez MA, de Soto-Del Río MLD, Gutiérrez RM, Chiu CY, Greninger AL, Contreras JF, López S, Arias CF, Isa P (2015) DNA microarray for detection of gastrointestinal viruses. J Clin Microbiol 53:136–145

44. Mori K, Hayashi Y, Akiba T, Nagano M, Tanaka T, Hosaka M, Nakama A, Kai A, Saito K, Shirasawa H (2013) Multiplex real-time PCR assays for the detection of group C rotavirus, astrovirus, and Subgenus F adenovirus in stool specimens. J Virol Methods 191:141–147

45. Ribeiro A, Ramalheira E, Cunha Â, Gomes N, Almeida A (2013) Incidence of rotavirus and adenovirus: detection by molecular and immunological methods in human faeces. J PURE Appl Microbiol 7:1505–1513

46. Takayama R, Hatakeyama N, Suzuki N, Yamamoto M, Hayashi T, Ikeda Y, Ikeda H, Nagano H, Ishida T, Tsutsumi H (2007) Quantification of adenovirus species B and C viremia by real-time PCR in adults and children undergoing stem cell transplantation. J Med Virol 79:278–284

47. van Maarseveen NM, Wessels E, de Brouwer CS, Vossen ACTM, Claas ECJ (2010) Diagnosis of viral gastroenteritis by simultaneous detection of Adenovirus group F, Astrovirus, Rotavirus group A, Norovirus genogroups I and II, and Sapovirus in two internally controlled multiplex real-time PCR assays. J Clin Virol 49:205–210

48. Oishi I, Yamazaki K, Kimoto T, Minekawa Y, Utagawa E, Yamazaki S, Inouye S, Grohmann G, Monroe S, Stine S, Carcamo C, Ando T, Glass R (1994) A large outbreak of acute gastro-

enteritis associated with astrovirus among students and teachers in Osaka, Japan. J Infect Dis 170:439–443

49. Clark B, McKendrick M (2004) A review of viral gastroenteritis. Curr Opin Infect Dis 17:461–469

50. Arraj A, Bohatier J, Aumeran C, Bailly JL, Laveran H, Traoré O (2008) An epidemiological study of enteric viruses in sewage with molecular characterization by RT-PCR and sequence analysis. J Water Health 6:351–358

51. Le Cann P, Ranarijaona S, Monpoeho S, Le Guyader F, Ferré V (2004) Quantification of human astroviruses in sewage using real-time RT-PCR. Res Microbiol 155:11–15

52. Ashley CR, Caul EO, Paver WK (1978) Astrovirus-associated gastroenteritis in children. J Clin Pathol 31:939–943

53. Guyader FSL, Saux J-CL, Ambert-Balay K, Krol J, Serais O, Parnaudeau S, Giraudon H, Delmas G, Pommepuy M, Pothier P, Atmar RL (2008) Aichi virus, norovirus, astrovirus, enterovirus, and rotavirus involved in clinical cases from a French oyster-related gastroenteritis outbreak. J Clin Microbiol 46:4011–4017

54. Logan C, O'Leary JJ, O'Sullivan N (2007) Real-time reverse transcription PCR detection of norovirus, sapovirus and astrovirus as causative agents of acute viral gastroenteritis. J Virol Methods 146:36–44

55. Kuan MM (1997) Detection and rapid differentiation of human enteroviruses following genomic amplification. J Clin Microbiol 35:2598–2601

56. Muir P, Kämmerer U, Korn K, Mulders MN, Pöyry T, Weissbrich B, Kandolf R, Cleator GM, van LAM (1998) Molecular typing of enteroviruses: current status and future requirements. Clin Microbiol Rev 11:202–227

57. CDC. Non-Polio Enterovirus | Home | Picornavirus. Available at: http://www.cdc.gov/non-polio-enterovirus/index.html. Accessed 24 May 2016

58. Tsai YL, Tran B, Sangermano LR, Palmer CJ (1994) Detection of poliovirus, hepatitis A virus, and rotavirus from sewage and ocean water by triplex reverse transcriptase PCR. Appl Environ Microbiol 60:2400–2407

59. Zhou J, Wang XC, Ji Z, Xu L, Yu Z (2015) Source identification of bacterial and viral pathogens and their survival/fading in the process of wastewater treatment, reclamation, and environmental reuse. World J Microbiol Biotechnol 31:109–120

60. Kitajima M, Hata A, Yamashita T, Haramoto E, Minagawa H, Katayama H (2013) Development of a reverse transcription-quantitative PCR system for detection and genotyping of Aichi viruses in clinical and environmental samples. Appl Environ Microbiol 79:3952–3958

61. Lago PM, Gary HE, Pérez LS, Cáceres V, Olivera JB, Puentes RP, Corredor MB, Jímenez P, Pallansch MA, Cruz RG (2003) Poliovirus detection in wastewater and stools following an immunization campaign in Havana, Cuba. Int J Epidemiol 32:772–777

62. Nijhuis M, van Maarseveen N, Schuurman R, Verkuijlen S, de Vos M, Hendriksen K, van Loon AM (2002) Rapid and sensitive routine detection of all members of the genus enterovirus in different clinical specimens by real-time PCR. J Clin Microbiol 40:3666–3670

63. Xiao X-L, He Y-Q, Yu Y-G, Yang H, Chen G, Li H-F, Zhang J-W, Liu D-M, Li X-F, Yang X-Q, Wu H (2008) Simultaneous detection of human enterovirus 71 and coxsackievirus A16 in clinical specimens by multiplex real-time PCR with an internal amplification control. Arch Virol 154:121–125

64. CDC. Hepatitis A Information | Division of Viral Hepatitis. Available at: http://www.cdc.gov/hepatitis/hav/. Accessed 24 May 2016

65. Wheeler C, Vogt TM, Armstrong GL, Vaughan G, Weltman A, Nainan OV, Dato V, Xia G, Waller K, Amon J, Lee TM, Highbaugh-Battle A, Hembree C, Evenson S, Ruta MA, Williams IT, Fiore AE, Bell BP (2005) An outbreak of hepatitis a associated with green onions. N Engl J Med 353:890–897

66. Grabow WOK (1997) Hepatitis viruses in water: update on risk and control. Water SA 23:379–386

67. Morace G, Aulicino FA, Angelozzi C, Costanzo L, Donadio F, Rapicetta M (2002) Microbial quality of wastewater: detection of hepatitis A virus by reverse transcriptase-polymerase chain reaction. J Appl Microbiol 92:828–836

68. Villar LM, De Paula VS, Diniz-Mendes L, Guimarães FR, Ferreira FFM, Shubo TC, Miagostovich MP, Lampe E, Gaspar AMC (2007) Molecular detection of hepatitis A virus in urban sewage in Rio de Janeiro, Brazil. Lett Appl Microbiol 45:168–173

69. Chitambar SD, Joshi MS, Sreenivasan MA, Arankalle VA (2001) Fecal shedding of hepatitis A virus in Indian patients with hepatitis A and in experimentally infected Rhesus monkey. Hepatol Res 19:237–246

70. Costafreda MI, Bosch A, Pintó RM (2006) Development, evaluation, and standardization of a real-time TaqMan reverse transcription-PCR assay for quantification of hepatitis A virus in clinical and shellfish samples. Appl Environ Microbiol 72:3846–3855

71. Sjogren MH, Tanno H, Fay O, Sileoni S, Cohen BD, Burke DS, Feighny RJ (1987) Hepatitis A virus in stool during clinical relapse. Ann Intern Med 106:221–226

72. Yotsuyanagi H, Koike K, Yasuda K, Moriya K, Shintani Y, Fujie H, Kurokawa K, Iino S (1996) Prolonged fecal excretion of hepatitis A virus in adult patients with hepatitis A as determined by polymerase chain reaction. Hepatology 24:10–13

73. Bonnet D, Kamar N, Izopet J, Alric L (2012) L'hépatite virale E: une maladie émergente. Rev Médecine Interne 33:328–334

74. CDC. Hepatitis E Information I Division of Viral Hepatitis. Available at: http://www.cdc.gov/hepatitis/hev/. Accessed 24 May 2016

75. Jothikumar N, Aparna K, Kamatchiammal S, Paulmurugan R, Saravanadevi S, Khanna P (1993) Detection of hepatitis E virus in raw and treated wastewater with the polymerase chain reaction. Appl Environ Microbiol 59:2558–2562

76. Clayson ET, Myint KSA, Snitbhan R, Vaughn DW, Innis BL, Chan L, Cheung P, Shrestha MP (1995) Viremia, Fecal Shedding, and IgM and IgG responses in patients with hepatitis E. J Infect Dis 172:927–933

77. Orrù G, Masia G, Orrù G, Romanò L, Piras V, Coppola RC (2004) Detection and quantitation of hepatitis E virus in human faeces by real-time quantitative PCR. J Virol Methods 118:77–82

78. Turkoglu S, Lazizi Y, Meng H, Kordosi A, Dubreuil P, Crescenzo B, Benjelloun S, Nordmann P, Pillot J (1996) Detection of hepatitis E virus RNA in stools and serum by reverse transcription-PCR. J Clin Microbiol 34:1568–1571

79. Fankhauser RL, Noel JS, Monroe SS, Ando T, Glass RI (1998) Molecular epidemiology of "Norwalk-like viruses" in outbreaks of gastroenteritis in the United States. J Infect Dis 178:1571–1578

80. Fankhauser RL, Monroe SS, Noel JS, Humphrey CD, Bresee JS, Parashar UD, Ando T, Glass RI (2002) Epidemiologic and molecular trends of "Norwalk-like viruses" associated with outbreaks of gastroenteritis in the United States. J Infect Dis 186:1–7

81. Mead PS, Slutsker L, Dietz V, McCaig LF, Bresee JS, Shapiro C, Griffin PM, Tauxe RV (1999) Food-related illness and death in the United States. Emerg Infect Dis 5:607–625

82. de Andrade Jda SR, Rocha MS, Carvalho-Costa FA, Fioretti JM, Xavier Mda PTP, Nunes ZMA, Cardoso J, Fialho AM, Leite JPG, Miagostovich MP (2014) Noroviruses associated with outbreaks of acute gastroenteritis in the State of Rio Grande do Sul, Brazil, 2004–2011. J Clin Virol 61:345–352

83. Kukkula M, Maunula L, Silvennoinen E, von Bonsdorff C-H (1999) Outbreak of viral gastroenteritis due to drinking water contaminated by Norwalk-like viruses. J Infect Dis 180:1771–1776

84. da Silva AK, Saux J-CL, Parnaudeau S, Pommepuy M, Elimelech M, Guyader FSL (2007) Evaluation of removal of noroviruses during wastewater treatment, using real-time reverse transcription-PCR: different behaviors of genogroups I and II. Appl Environ Microbiol 73:7891–7897

85. Hewitt J, Greening GE, Leonard M, Lewis GD (2013) Evaluation of human adenovirus and human polyomavirus as indicators of human sewage contamination in the aquatic environment. Water Res 47:6750–6761

86. Atmar RL, Opekun AR, Gilger MA, Estes MK, Crawford SE, Neill FH, Graham DY (2008) Norwalk virus shedding after experimental human infection. Emerg Infect Dis 14:1553–1557
87. Fumian TM, Justino MCA, Mascarenhas JDP, Reymão TKA, Abreu E, Soares L, Linhares AC, Gabbay YB (2013) Quantitative and molecular analysis of noroviruses RNA in blood from children hospitalized for acute gastroenteritis in Belém, Brazil. J Clin Virol 58:31–35
88. Schvoerer E, Bonnet F, Dubois V, Cazaux G, Serceau R, Fleury HJ, Lafon ME (2000) PCR detection of human enteric viruses in bathing areas, waste waters and human stools in southwestern France. Res Microbiol 151:693–701
89. CDC. Rotavirus | Home | Gastroenteritis. Available at: http://www.cdc.gov/rotavirus/. Accessed 24 May 2016
90. Li D, Gu AZ, Zeng S-Y, Yang W, He M, Shi H-C (2011) Monitoring and evaluation of infectious rotaviruses in various wastewater effluents and receiving waters revealed correlation and seasonal pattern of occurrences. J Appl Microbiol 110:1129–1137
91. Meleg E, Bányai K, Martella V, Jiang B, Kocsis B, Kisfali P, Melegh B, Szűcs G (2008) Detection and quantification of group C rotaviruses in communal sewage. Appl Environ Microbiol 74:3394–3399
92. Fumian TM, Gagliardi Leite JP, Rose TL, Prado T, Miagostovich MP (2011) One year environmental surveillance of rotavirus specie A (RVA) genotypes in circulation after the introduction of the Rotarix® vaccine in Rio de Janeiro, Brazil. Water Res 45:5755–5763
93. Baggi F, Peduzzi R (2000) Genotyping of rotaviruses in environmental water and stool samples in Southern Switzerland by nucleotide sequence analysis of 189 base pairs at the 5' end of the VP7 gene. J Clin Microbiol 38:3681–3685
94. Gouvea V, Glass RI, Woods P, Taniguchi K, Clark HF, Forrester B, Fang ZY (1990) Polymerase chain reaction amplification and typing of rotavirus nucleic acid from stool specimens. J Clin Microbiol 28:276–282
95. Mukhopadhya I, Sarkar R, Menon VK, Babji S, Paul A, Rajendran P, Sowmyanarayanan TV, Moses PD, Iturriza-Gomara M, Gray JJ, Kang G (2013) Rotavirus shedding in symptomatic and asymptomatic children using reverse transcription-quantitative PCR. J Med Virol 85:1661–1668
96. Ambert-Balay K, Lorrot M, Bon F, Giraudon H, Kaplon J, Wolfer M, Lebon P, Gendrel D, Pothier P (2008) Prevalence and genetic diversity of Aichi virus strains in stool samples from community and hospitalized patients. J Clin Microbiol 46:1252–1258
97. Haramoto E, Kitajima M, Otagiri M (2013) Development of a reverse transcription-quantitative PCR assay for detection of salivirus/klassevirus. Appl Environ Microbiol 79:3529–3532
98. Oka T, Katayama K, Hansman GS, Kageyama T, Ogawa S, Wu F, White PA, Takeda N (2006) Detection of human sapovirus by real-time reverse transcription-polymerase chain reaction. J Med Virol 78:1347–1353
99. Siebrasse EA, Reyes A, Lim ES, Zhao G, Mkakosya RS, Manary MJ, Gordon JI, Wang D (2012) Identification of MW polyomavirus, a novel polyomavirus in human stool. J Virol 86:10321–10326
100. Hino S, Miyata H (2006) Torque teno virus (TTV): current status. Rev Med Virol 17:45–57
101. Lin K, Marr LC (2017) Aerosolization of Ebola virus surrogates in wastewater systems. Environ Sci Technol 51:2669–2675
102. Cotruvo JA, Dufour A, Rees G, Bartram J, Carr R, Cliver DO, Craun GF, Fayer R, Gannon VP (2004) Waterborne zoonoses. IWA Publishing, London, UK
103. Gundy PM, Gerba CP, Pepper IL (2008) Survival of coronaviruses in water and wastewater. Food Environ Virol 1:10–14
104. Hu Y, Lu S, Song Z, Wang W, Hao P, Li J, Zhang X, Yen H-L, Shi B, Li T, Guan W, Xu L, Liu Y, Wang S, Zhang X, Tian D, Zhu Z, He J, Huang K, Chen H, Zheng L, Li X, Ping J, Kang B, Xi X, Zha L, Li Y, Zhang Z, Peiris M, Yuan Z (2013) Association between adverse clinical outcome in human disease caused by novel influenza A H7N9 virus and sustained viral shedding and emergence of antiviral resistance. The Lancet 381:2273–2279

105. Lee N, Chan PK, Wong CK, Wong K-T, Choi K-W, Joynt GM, Lam P, Chan MC, Wong BC, Lui GC, Sin WW, Wong RY, Lam W-Y, Yeung AC, Leung T-F, So H-Y, Yu AW, Sung JJ, Hui DS (2011) Viral clearance and inflammatory response patterns in adults hospitalized for pandemic 2009 influenza A(H1N1) virus pneumonia. Antivir Ther 16:237–247

106. To KKW, Chan K-H, Li IWS, Tsang T-Y, Tse H, Chan JFW, Hung IFN, Lai S-T, Leung C-W, Kwan Y-W, Lau Y-L, Ng T-K, Cheng VCC, Peiris JSM, Yuen K-Y (2010) Viral load in patients infected with pandemic H1N1 2009 influenza A virus. J Med Virol 82:1–7

107. Heijnen L, Medema G (2011) Surveillance of influenza A and the pandemic influenza A (H1N1) 2009 in sewage and surface water in the Netherlands. J Water Health 9:434–442

108. Gourinat A-C, O'Connor O, Calvez E, Goarant C, Dupont-Rouzeyrol M (2015) Detection of Zika virus in urine. Emerg Infect Dis 21:84–86

109. Barzon L, Pacenti M, Franchin E, Pagni S, Martello T, Cattai M, Cusinato R, Palù G (2013) Excretion of West Nile virus in urine during acute infection. J Infect Dis 208:1086–1092

110. Tonry JH, Brown CB, Cropp CB, Co JKG, Bennett SN, Nerurkar VR, Kuberski T, Gubler DJ (2005) West Nile Virus detection in urine. Emerg Infect Dis 11:1294–1296

111. Hirayama T, Mizuno Y, Takeshita N, Kotaki A, Tajima S, Omatsu T, Sano K, Kurane I, Takasaki T (2012) Detection of dengue virus genome in urine by real-time reverse transcriptase PCR: a laboratory diagnostic method useful after disappearance of the genome in serum. J Clin Microbiol 50:2047–2052

112. Mizuno Y, Kotaki A, Harada F, Tajima S, Kurane I, Takasaki T (2007) Confirmation of dengue virus infection by detection of dengue virus type 1 genome in urine and saliva but not in plasma. Trans R Soc Trop Med Hyg 101:738–739

113. Poloni TR, Oliveira AS, Alfonso HL, Galvao LR, Amarilla AA, Poloni DF, Figueiredo LT, Aquino VH (2010) Detection of dengue virus in saliva and urine by real time RT-PCR. Virol J 7:22

114. Nielsen ACY, Gyhrs ML, Nielsen LP, Pedersen C, Böttiger B (2013) Gastroenteritis and the novel picornaviruses Aichi virus, cosavirus, saffold virus, and salivirus in young children. J Clin Virol 57:239–242

115. Vanchiere JA, Nicome RK, Greer JM, Demmler GJ, Butel JS (2005) Frequent detection of polyomaviruses in stool samples from hospitalized children. J Infect Dis 192:658–664

116. Vanchiere JA, Abudayyeh S, Copeland CM, Lu LB, Graham DY, Butel JS (2009) Polyomavirus shedding in the stool of healthy adults. J Clin Microbiol 47:2388–2391

117. Li L, Victoria J, Kapoor A, Blinkova O, Wang C, Babrzadeh F, Mason CJ, Pandey P, Triki H, Bahri O, Oderinde BS, Baba MM, Bukbuk DN, Besser JM, Bartkus JM, Delwart EL (2009) A novel picornavirus associated with gastroenteritis. J Virol 83:12002–12006

118. Chan MCW, Sung JJY, Lam RKY, Chan PKS, Lai RWM, Leung WK (2006) Sapovirus detection by quantitative real-time RT-PCR in clinical stool specimens. J Virol Methods 134:146–153

119. Ross RS, Viazov S, Runde V, Schaefer UW, Roggendorf M (1999) Detection of TT virus DNA in specimens other than blood. J Clin Virol 13:181–184

120. Chan KH, Poon LL, Cheng VC, Guan Y, Hung IF, Kong J, Yam LY, Seto WH, Yuen KY, Peiris JS (2004) Detection of SARS coronavirus in patients with suspected SARS. Emerg Infect Dis 10:294–299

121. Poon LLM, Chan KH, Wong OK, Cheung TKW, Ng I, Zheng B, Seto WH, Yuen KY, Guan Y, Peiris JSM (2004) Detection of SARS coronavirus in patients with severe acute respiratory syndrome by conventional and real-time quantitative reverse transcription-PCR assays. Clin Chem 50:67–72

122. Vester D, Lagoda A, Hoffmann D, Seitz C, Heldt S, Bettenbrock K, Genzel Y, Reichl U (2010) Real-time RT-qPCR assay for the analysis of human influenza A virus transcription and replication dynamics. J Virol Methods 168:63–71

123. dos Santos HWG, Poloni TRRS, Souza KP, Muller VDM, Tremeschin F, Nali LC, Fantinatti LR, Amarilla AA, Castro HLA, Nunes MR, Casseb SM, Vasconcelos PF, Badra SJ, Figueiredo LTM, Aquino VH (2008) A simple one-step real-time RT-PCR for diagnosis of dengue virus infection. J Med Virol 80:1426–1433

124. Linke S, Ellerbrok H, Niedrig M, Nitsche A, Pauli G (2007) Detection of West Nile virus lineages 1 and 2 by real-time PCR. J Virol Methods 146:355–358
125. Faye O, Faye O, Diallo D, Diallo M, Weidmann M, Sall A (2013) Quantitative real-time PCR detection of Zika virus and evaluation with field-caught mosquitoes. Virol J 10:311
126. Domingo C, Yactayo S, Agbenu E, Demanou M, Schulz AR, Daskalow K, Niedrig M (2011) Detection of yellow fever 17D genome in urine. J Clin Microbiol 49:760–762
127. Bae H-G, Nitsche A, Teichmann A, Biel SS, Niedrig M (2003) Detection of yellow fever virus: a comparison of quantitative real-time PCR and plaque assay. J Virol Methods 110:185–191
128. Orenstein EW, Fang ZY, Xu J, Liu C, Shen K, Qian Y, Jiang B, Kilgore PE, Glass RI (2007) The epidemiology and burden of rotavirus in China: a review of the literature from 1983 to 2005. Vaccine 25:406–413
129. Bofill-Mas S, Pina S, Girones R (2000) Documenting the epidemiologic patterns of poly-omaviruses in human populations by studying their presence in urban sewage. Appl Environ Microbiol 66:238–245
130. Clemente-Casares P, Pina S, Buti M, Jardi R, Martín M, Bofill-Mas S, Girones R (2003) Hepatitis E virus epidemiology in industrialized countries. Emerg Infect Dis 9:448–454
131. Carducci A, Verani M, Battistini R, Pizzi F, Rovini E, Andreoli E, Casini B (2006) Epidemiological surveillance of human enteric viruses by monitoring of different environ-mental matrices. Water Sci Technol 54:239–244
132. O'Brien E, Nakyazze J, Wu H, Kiwanuka N, Cunningham W, Kaneene JB, Xagoraraki I (2017b) Viral diversity and abundance in polluted waters in Kampala, Uganda. Water Res 127:41–49
133. Bibby K, Peccia J (2013) Identification of viral pathogen diversity in sewage sludge by metagenome analysis. Environ Sci Technol 47:1945–1951
134. Tamaki H, Zhang R, Angly FE, Nakamura S, Hong P-Y, Yasunaga T, Kamagata Y, Liu W-T (2012) Metagenomic analysis of DNA viruses in a wastewater treatment plant in tropical climate. Environ Microbiol 14:441–452
135. Aw TG, Howe A, Rose JB (2014) Metagenomic approaches for direct and cell culture evalu-ation of the virological quality of wastewater. J Virol Methods 210:15–21
136. Cantalupo PG, Calgua B, Zhao G, Hundesa A, Wier AD, Katz JP, Grabe M, Hendrix RW, Girones R, Wang D, Pipas JM (2011) Raw sewage harbors diverse viral populations. MBio 2:e00180–e00111
137. Bibby K, Viau E, Peccia J (2011) Viral metagenome analysis to guide human pathogen moni-toring in environmental samples. Lett Appl Microbiol 52:386–392
138. Finkbeiner SR, Allred AF, Tarr PI, Klein EJ, Kirkwood CD, Wang D (2008) Metagenomic analysis of human diarrhea: viral detection and discovery. PLoS Pathog 4:e1000011
139. Victoria JG, Kapoor A, Li L, Blinkova O, Slikas B, Wang C, Naeem A, Zaidi S, Delwart E (2009) Metagenomic analyses of viruses in stool samples from children with acute flaccid paralysis. J Virol 83:4642–4651
140. Lai FY, Anuj S, Bruno R, Carter S, Gartner C, Hall W, Kirkbride KP, Mueller JF, O'Brien JW, Prichard J, Thai PK, Ort C (2015) Systematic and day-to-day effects of chemical-derived population estimates on wastewater-based drug epidemiology. Environ Sci Technol 49:999–1008
141. O'Brien JW, Thai PK, Eaglesham G, Ort C, Scheidegger A, Carter S, Lai FY, Mueller JF (2014) A model to estimate the population contributing to the wastewater using samples col-lected on census day. Environ Sci Technol 48:517–525
142. Chiaia AC, Banta-Green C, Field J (2008) Eliminating solid phase extraction with large-volume injection LC/MS/MS: analysis of illicit and legal drugs and human urine indicators in US wastewaters. Environ Sci Technol 42:8841–8848
143. Daughton CG (2012) Real-time estimation of small-area populations with human biomarkers in sewage. Sci Total Environ 414:6–21
144. Baker DR, Kasprzyk-Hordern B (2011) Multi-residue analysis of drugs of abuse in wastewa-ter and surface water by solid-phase extraction and liquid chromatography–positive electro-spray ionisation tandem mass spectrometry. J Chromatogr 1218:1620–1631

145. Chen C, Kostakis C, Gerber JP, Tscharke BJ, Irvine RJ, White JM (2014) Towards finding a population biomarker for wastewater epidemiology studies. Sci Total Environ 487:621–628

146. Been F, Rossi L, Ort C, Rudaz S, Delémont O, Esseiva P (2014) Population normalization with ammonium in wastewater-based epidemiology: application to illicit drug monitoring. Environ Sci Technol 48:8162–8169

147. Edwards AC, Withers PJA (2007) Linking phosphorus sources to impacts in different types of water body. Soil Use Manag 23:133–143

148. James E, Kleinman P, Veith T, Stedman R, Sharpley A (2007) Phosphorus contributions from pastured dairy cattle to streams of the Cannonsville Watershed, New York. J Soil Water Conserv 62:40–47

149. Chang H, Wan Y, Wu S, Fan Z, Hu J (2011) Occurrence of androgens and progestogens in wastewater treatment plants and receiving river waters: comparison to estrogens. Water Res 45:732–740

150. Chang H, Hu J, Shao B (2007) Occurrence of natural and synthetic glucocorticoids in sewage treatment plants and receiving river waters. Environ Sci Technol 41:3462–3468

151. Cheng H-Y, Huang Y-C, Yen T-Y, Hsia S-H, Hsieh Y-C, Li C-C, Chang L-Y, Huang L-M (2014) The correlation between the presence of viremia and clinical severity in patients with enterovirus 71 infection: a multi-center cohort study. BMC Infect Dis 14:417

152. Bosch A, Pinto RM, Abad FX (2006) Survival and transport of enteric viruses in the environment. In: Goyal S.M. (eds) Viruses in Foods. Food Microbiology and Food Safety. Springer, Boston, MA

153. Xagoraraki I, Yin Z, Svambayev Z (2014) Fate of viruses in water systems. J Environ Eng 140:04014020

Irene Xagoraraki is an Associate Professor of Environmental Engineering in the Department of Civil and Environmental Engineering at Michigan State University. Her bachelor's degree (1993) in environmental science is from the University of the Aegean in Greece. Her bachelor's thesis focused on landfill biogas production and recovery, under the guidance of Prof. Halvadakis. She earned her MS (1995) and PhD (2001) degrees in Civil and Environmental Engineering from the University of Wisconsin-Madison. Her MS degree, under the guidance of Prof. Berthouex, focused on pumping station reliability analysis for the Madison metropolitan sewerage district. Her PhD advisor was Prof. Harrington, and her dissertation focused on coagulation and sedimentation of *Cryptosporidium parvum* in water systems. Between 2001 and 2005, she held a postdoctoral position at the University of Wisconsin-Madison where she worked on multiple pathogen removal research projects for drinking water treatment systems. She joined the faculty of Michigan State University in 2006. The focus of her teaching is on environmental engineering, water quality and public health.

Dr. Xagoraraki's research program at Michigan State University is focused on water-quality engineering, emphasizing protection of public health and prevention of waterborne disease. In particular, she is interested in microbial contaminants (such as human viruses, zoonotic viruses, antibiotic-resistant bacteria) and their detection, occurrence, fate, removal, inactivation, and associated risk, in water systems. Her research projects are funded by National Science Foundation, US Environmental Protection Agency, US Geological Survey, US Department of Homeland Security, Water Research Foundation, and Water Environmental Research Foundation. Her latest research project, funded by the National Science Foundation, explores wastewater-based epidemiology methods for early detection and prediction of water-related viral outbreaks. The basic principles of this methodology are described in the article presented in this book. Dr. Xagoraraki is also involved in multiple international projects in Uganda, Mexico, France, and the Republic of Georgia. Dr. Xagoraraki loves to be surrounded by water. She enjoys the beautiful shores of the Great Lakes in Michigan, but to find inspiration and reconnect with her origins, she visits the incredible Aegean Sea in the island of Crete, Greece, where she was born and raised.

Evan O'Brien earned his PhD in Environmental Engineering at Michigan State University in 2018. He earned his MS degree (2014) in Environmental Engineering from Michigan State University and his BS degree (2011) in Nuclear Engineering with a minor in Environmental Engineering from The Pennsylvania State University. His research work involves the investigation and characterization of viral diversity in wastewater and surface waters. His first project utilized next-generation sequencing and metagenomic analyses to compare viral diversity in wastewater effluents between conventional activated sludge and membrane bioreactor wastewater treatment plants. His subsequent project employed these methods in addition to RT-qPCR to investigate viral abundance and diversity in wastewater and its impact on surface water in Uganda. His interest in the field of water resources arises from a lifelong appreciation of wilderness and the outdoors, cultivated while working as a camp counselor in northern Wisconsin. He hopes that his future career will work to ensure public and environmental health, so that these natural resources can be enjoyed for generations to come.

Chapter 6
Urine Source Separation for Global Nutrient Management

Tove A. Larsen

Abstract The sewer-based paradigm for wastewater management at the global scale is not successful neither from a humanitarian nor from an environmental perspective. The systems are too expensive for the largest part of the global population. Source separation and resource recovery offer an alternative for sanitation and water pollution control. This chapter illustrates the importance but also the challenges of urine source separation for efficient nutrient removal and recovery.

6.1 Introduction

Urine source separation for nutrient recovery can be compared to the well-known source separation for solid waste, but it was only adopted by the scientific community in the 1990s [14–16, 19, 24, 25].

Urine separation would be a disruptive innovation in the entire system of urban wastewater management, leading to larger investments at the household level but smaller investments in municipal infrastructure. This chapter argues that the innovation is not only beneficial but may also be essential in order to achieve sufficient nutrient removal from wastewater in most areas of the world. It also explores some of the challenges as well as the technical opportunities for introducing such a radical change to the urban water cycle and gives a few examples of urine separation as an addition to rural wastewater treatment.

T. A. Larsen (✉)
Eawag, Swiss Federal Institute of Aquatic Science and Technology, Department of Urban Water Management Überlandstrasse 133, Dübendorf, Switzerland
e-mail: tove.larsen@eawag.ch

99
D. J. O'Bannon (ed.), *Women in Water Quality*, Women in Engineering
and Science, https://doi.org/10.1007/978-3-030-17819-2_6

6.1.1   Why Urine Source Separation?

Urine contains the majority of the water-relevant nutrients nitrogen (N) and phosphorus (P) excreted from human metabolism: nearly 90% of N and nearly 60% of P, as well as large amounts of potassium (K) [24]. With a present world population of 7.5 billion, these nutrient contributions amount to roughly 25 Mt/year of nitrogen, 1.8 Mt/year of phosphorus, and 5.5 Mt/year of potassium. These nutrients have a value, if recovered, of more than US$20 billion/year on the global market, even at the relatively low fertilizer prices experienced in 2014 [54], and for the production of this amount of ammonia for agricultural use, about 1100 PWh/year of primary energy is required [33]. Further it should be noted that local fertilizer prices may be up to a factor of 2–3 higher than global market prices, especially in landlocked countries with poor supply networks [13].

Fertilizer costs are high, but they are dwarfed by the potential costs of protecting the environment from the nutrients through sewer-based wastewater treatment. Nutrient elimination is part of urban water management in order to protect surface water against eutrophication. In industrial countries, the costs of nutrient elimination are estimated at US$20–40/cap/year [34]. If we equip the present global population of 7.5 billion with this nutrient-removal technology as an addition to basic wastewater treatment plants, the total costs will be US$150–300 billion/year—about 10 times higher than the market value of the nutrients.

In reality, the costs will be much higher. The estimated costs of nutrient elimination cited above imply that sewers and basic treatment plants (without nutrient elimination) are already available. For the 70% of the world population, which is presently not connected to any sewer system [46], the global costs would be about seven times higher to establish collection systems and basic treatment plants [34]. The total global costs of introducing conventional nutrient elimination for the entire global population would thus be in the order of size of a trillion US$/year. The advantage of urine separation technology is that a large part of the huge amount of nutrients in urine can be collected without any initial investments in a sewer system.

Existing biological treatment plants would not require any upgrading to advanced nutrient removal if urine separation is implemented. Without N and P from urine, domestic wastewater would have a balanced C:N:P ratio, meaning that a simple biological wastewater treatment plant with a short sludge retention time (SRT) could reach excellent N and P emission standards [24]. In a modeling study for a specific catchment, Wilsenach and van Loosdrecht [51] showed that with biological aerobic treatment at an SRT of 0.8 day (as compared to at least 12 days and often much more for a typical nitrogen-removing plant), the process would become N-limited at 90% urine separation. The effluent P concentration would be around 1 gm^{-3}, which is a typical effluent requirement, but it would only take a very small amount of precipitant to reduce this value further if needed for sensitive aquatic environments. Since the tank size of the biological unit of a wastewater treatment plant is proportional to SRT, such a plant could be much smaller and cheaper.

6.1.2   Challenges

There are good reasons to introduce urine source separation as part of mainstream wastewater management. There are, however, a number of reasons why urine separation is still only implemented in simple settings, primarily in rural settings of low-income countries. At present, there is neither technology nor institutions for introducing widespread urine source separation in an urban environment. Furthermore, transport of urine is difficult, calling for decentralized or even on-site technologies, which are traditionally not in the repertoire of wastewater management authorities. Technologies, including small scale, for urine treatment are in development and so are urine-separating flush toilets (NoMix toilets). The most difficult hurdle may be the attitude of stakeholders and decision-makers: Why change a successful system like advanced nutrient removal when it is not absolutely necessary? In order to understand why this may still be rewarding, this chapter will start with a review of the problems and challenges of conventional advanced nutrient removal.

6.2   Global Nutrient Balances

Advanced nutrient elimination at treatment plants is a powerful technology, which has allowed maintaining or recovering the quality of aquatic ecosystems in many countries. However, in even more countries, it has not been possible to introduce this technology although it has been available for nearly half a century. Additionally, there are other disadvantages which could be mitigated with innovative technology: issues of energy consumption and climate change.

6.2.1   The Role of Wastewater Nutrients in Eutrophication of Aquatic Ecosystems

The role of nitrogen and phosphorus in eutrophication of aquatic ecosystems is well-established. As a general rule of thumb, phosphorus causes eutrophication in lakes, and nitrogen causes eutrophication in coastal areas [18]. Normally, agriculture is considered the main source of N and P, but urban areas are also contributing significantly to nutrient loads [52]. The important role of larger cities in hypoxia has been shown, e.g., by Lajaunie-Salla et al. [22], and as argued by McCrackin et al. [37], even where the emissions are dominated by agriculture, a combined effort including nutrient removal at treatment plants is often required to solve the problem. Effective nutrient removal from wastewater is thus a very powerful tool for improving the quality of aquatic ecosystems in the vicinity of cities.

6.2.2 Advanced Nutrient Elimination on a Global Scale

Globally, wastewater treatment plants remove only 20% of P and 10% of N contained in wastewater (Fig. 6.1). The reason is that low- and middle-income countries have not invested in nutrient-eliminating wastewater treatment plants [47]. The same figure projects that the amounts of N and P emitted to the environment will increase in the future: it is not possible to construct infrastructure rapidly enough to cope with the increased nutrient emissions of a growing world population.

Most of the nutrients are and will be emitted from people who are not connected to wastewater treatment plants. Where people are connected, removal rate is increasing, but population growth and better nutrition are expected to outweigh this improvement. Data and model predictions are compiled from Bouwman et al. [5] and Van Drecht et al. [46]. Model predictions for 2030 and 2050 are average values from four generally accepted socio-economic models for the future.

6.2.3 Climate Relevance of Wastewater Nitrogen Transformation

Nitrogen transformation at wastewater treatment plants is climate-relevant in several ways: nitrous oxide (an important climate gas) is emitted from nitrifying treatment plants, energy is consumed in the nitrification process, and production of energy-rich sludge is reduced due to the long solid retention time (SRT) required for conventional nitrogen transformations. Nitrogen removal (transformation to N_2) also means a loss of chemical energy embedded in reactive nitrogen through nitrogen fixation (transformation of N_2 to ammonia, either biologically or chemically).

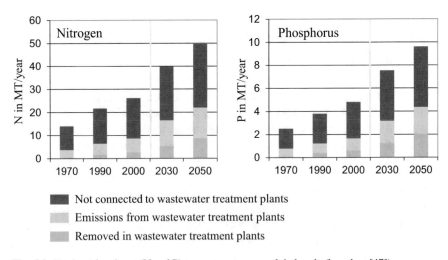

Fig. 6.1 Total nutrient input (N and P) to wastewater on a global scale (based on [47])

It can be shown that nitrogen recovery from urine holds the highest potential for climate-friendly wastewater management, especially if N_2O emissions can be avoided, e.g., if recovery would take place without ammonia oxidation [23].

6.3   Nutrient Recovery

Steffen et al. [41] identified the major areas of global environmental risk, using the notion of "planetary boundaries." The idea is that within these boundaries, the risk of destabilizing the Earth is low. According to the authors, the emission of N and P already surpasses the corresponding "safe" boundary, primarily due to eutrophication of the oceans on a global scale. In earlier similar work, Rockstrom et al. [39] pleaded for a strongly reduced production of reactive nitrogen (all nitrogen species except N_2), in order to keep the amount of reactive nitrogen within the boundaries identified as "safe." At the same time, large amounts of nitrogen are essential to feed a growing population; nitrogen recovery from wastewater is thus a logical path to follow.

6.3.1   Nutrient Recycling from Conventional Wastewater Treatment Plants

The most direct way of recovering nutrients from wastewater is direct use of wastewater for irrigation. This technology was used historically, e.g., in Paris from 1872, but already from the beginning, the high demand for space made it impossible to treat the entire wastewater from Paris [48]. The technology is still in use at places where enough agricultural land is available and where consumers tolerate the risk of anthropogenic contamination.

All types of phosphorus removal on treatment plants transfer phosphorus to the sludge fraction. Sludge is frequently spread to agricultural land, primarily to save costs for sludge disposal, but also in order to recycle phosphorus. Depending on the processing technology, phosphorus from sludge may be bioavailable [20], but in some European countries, consumers are opposed to the use of sewage sludge on agricultural land due to the fear of inorganic, organic, and infectious contamination. In Switzerland, this fear resulted in a legal ban of wastewater application [28], whereas in Sweden, the use was strongly reduced because the large food retailers do not accept food produced in areas fertilized with sewage sludge [1].

Twenty percent of wastewater nitrogen can be recovered from digester supernatant through stripping [3], but this technology has not yet been broadly implemented. Within the existing system, it is difficult to recover more than 20%. In principle, ion exchange with zeolites [42, 50] and adsorption on activated carbon [9] from wastewater are possible, but this is not efficient. More experience is available for the uptake of nitrogen from municipal wastewater in biomass. Effective uptake of N can be achieved with water hyacinths or duckweed [55], but for uptake with less space requirements, algal ponds are considered more suitable. However, even for high-rate algal ponds (HRAP), space demand will be around 10 m^2/person (based on loading information in Craggs et al. [8]), resulting in a large space requirement of 50 ha for a medium-sized town of 50,000 inhabitants, which is unlikely to be implemented.

The difficulties of recovering nitrogen from diluted wastewater is one of the most important reasons for discussing nutrient recovery from toilet waste. Urine contains around 90% of the nitrogen from human metabolism; therefore, urine separation is the most obvious technology for efficient nitrogen recovery.

Direct use of urine and even combined toilet waste are traditional technologies [6], which obviously have a large potential to recycle nutrients from human metabolism. New sanitization methods have been developed and tested: e.g., prolonged storage for urine [17] and urea treatment for combined toilet waste, aka blackwater [11]. As shown by a number of Life Cycle Assessment (LCA) studies, direct recycling of urine and blackwater is highly beneficial from an environmental point of view, as long as ammonia emissions are kept to a minimum (through proper storage and spreading technologies) and transport distances to agricultural land are not excessive (see, e.g., Tidåker et al. [43]). For urban areas, transport of urine to agriculture is not practical, and processing of urine is therefore necessary.

6.4 Urine Separation in an Urban Environment

Recovery of nutrients from urine in urban settings places high demands on technology. Maurer et al. [35] state that the process engineering goals are primarily volume reduction, stabilization, sanitization, and removal of micropollutants. Volume is certainly the most important issue for transfer of urine from urban areas to agriculture. Ledezma et al. [27] report findings that in Australia, the transport of urine from urban areas to agriculture would be more expensive than the costs of the equivalent fertilizers. The volume of urine, at 1.5 L/person/day, is on the same order of magnitude as domestic solid waste. Collection in a city would thus mean doubling the transport requirements for domestic waste—a huge effort with large financial consequences. Stabilization with respect to odor and ammonia emissions is equally important: weekly collection of unstabilized urine in an urban environment would probably be possible, but at a high price. Sanitization is necessary mainly due to cross-contamination with fecal pathogens [17]. Finally, micropollutants in urine are important: about two-thirds of the pharmaceuticals excreted from human metabolism are contained in urine [30]. As shown by Bischel et al. [2], high concentrations of pharmaceuticals may be found in urine from areas with a high prevalence of HIV. Although those are the more polar (water-soluble) ones and therefore less critical than the lipid-soluble ones contained in feces, urine still contains half of the ecotoxicologic potential from human excretions [31].

6.4.1   Technical Options for Nutrient Recovery from Urine

Urine is essentially a highly concentrated nutrient solution with a few annoying properties [44]. The main problems are odor and a high pH value, which causes a high concentration of gaseous ammonia and thus can lead to large nitrogen emissions during transport and handling of urine. For more than 10 years, a small number of researchers have developed different biological, physical-chemical, and electrochemical approaches in order to reduce the volume, stabilize the solution (i.e., reduce odor and pH), and eventually produce a hygienic fertilizer, possibly without micropollutants. Maurer et al. [35] documented the first attempts, and in the meantime, many other technologies are under development. One of the most-developed technologies is the stabilization of ammonia by partial nitrification (decreasing the pH value and thus pushing the equilibrium toward nonvolatile ammonium) followed by distillation for volume reduction [12].

The choice of urine treatment technology at each site will depend on a large number of local factors, which will not be discussed here. McConville et al. [36] present a suitable methodology for system designs.

6.4.2   Acceptance of Urine Source Separation

A common criticism of urine source separation concerns public acceptance. Many people anticipate that users will not accept NoMix toilets and that farmers will not want to use a urine-based fertilizer. Whereas it is true that users are critical toward NoMix toilets (shown in Fig. 6.2), especially when they have them at home, this has

Fig. 6.2 The NoMix toilet, which has been in use at Eawag since 2006. Urine is collected at the front; the rest of the toilet functions as any conventional toilet. A special mechanism ensures that only a little flush water finds its way into the urine tank © Eawag

more to do with the toilet technology on the market than with a general lack of acceptance. In a review, Lienert and Larsen [29] found that in Europe, about 80% of people who have experienced the present NoMix toilets are in favor of the technology despite the daily problems with the toilets. People living with the toilets at home tend to be more critical than those experiencing NoMix toilets only at the workplace. Consumers as well as farmers [38] would also accept a urine-based fertilizer if they could be convinced that this fertilizer would be safe: safety being associated with hygiene and the absence of pharmaceuticals from human metabolism.

In areas where urine-diverting dry toilets (UDDTs) are applied, satisfaction with existing toilet technology is poor. In the municipality of eThekwini, South Africa, Roma et al. [40] found a very low satisfaction with the toilets after 10 years of implementation. New and better technologies are needed for UDDT acceptance.

6.5   Future Research and Implementation Projects

Researchers and some practitioners have been discussing urine separation for about 20 years, but still there are very few full-scale implementations. Although a period of 20 years is not a long development time for a highly disruptive technology like urine source separation, it is valuable to reflect on the obstacles and the way forward.

There are many places where urine separation would make sense and where one could develop good transition scenarios. Three of those situations will be discussed here.

Urine separation could avoid expensive extensions of wastewater treatment plants in growing cities [4], a solution which has also been investigated in a master study at the water utility, the Greater Paris Sanitation Authority (SIAAP) in Paris [7]. Paris will grow by 10% over the next 10 years and the treatment plants have no capacity to cope with the extra nitrogen. The introduction of targets for nitrogen emissions in cities that previously had none results in the same situation, as exemplified by the city of Hong Kong. In this city, on-site nitrification of urine followed by denitrification in the sewers is considered as an alternative to the upgrading of existing treatment plants to nitrifying-denitrifying plants [56].

Another possible transition scenario is on-site wastewater treatment in rural areas of high-income countries, which may help mature the technology for later urban implementation. Wood et al. [53] have shown that urine separation would be a low-cost, very effective add-on to existing septic tanks. In a similar situation, urine source separation is already implemented in Sweden as a powerful technology for preventing nutrient pollution from rural areas where only simple wastewater treatment is available [49].

The third possible situation is again found in Hong Kong, where seawater is used for toilet flushing due to water scarcity. Seawater provides high sulfate concentrations in the sewers and anaerobic conditions therefore lead to sulfide production causing corrosion [10]. On-site nitrification followed by denitrification in the sewers would not only solve the problem of treatment plants as discussed above but also prevent sulfide production in the sewers. Microorganisms prefer nitrate over sulfate as an electron acceptor, producing inert N_2 gas instead of hydrogen sulfide (H_2S). Hong Kong foresees recovery of phosphorus as struvite (magnesium ammonium phosphate) with seawater contributing the magnesium source, followed by on-site full nitrification and in-sewer denitrification [32].

Today, full-scale urine separation is primarily implemented in rural areas of low-income countries where it makes dry toilets less smelly and reduces logistic costs because urine is considered less of a hygienic risk and is, therefore, just infiltrated. The largest implementation of this type of urine separation is the eThekwini water utility in Durban, South Africa, where more than 75,000 urine-separating dry toilets were in use in 2013 [45]. A large research project, financed by the Bill & Melinda Gates Foundation (www.eawag.ch/vuna), investigated the possibility of introducing collection of urine followed by nutrient recovery. The reason for urine collection is, on one hand, the protection of drinking water resources and, on the other hand, the fertilizer production from urine.

The largest hurdle for further developments of urine source separation in low-income as well as in high-income countries is the lack of good urine-separating toilets, be it in the dry variant as UDDTs or in the flushed variant as NoMix toilets. In low-income countries, cheap urine-separating solutions already exist, but the aspirations of low-income users are for conventional flush toilets, not for better UDDTs [40]. It has been shown that these hurdles can be overcome by combining UDDTs with a separate water cycle [21, 26]: it will be necessary to bring the costs of those improved UDDTs down to a level where they are economically attractive for the target population. For high-income countries, the design demands from consumers for NoMix-type toilets are requiring ceramic producers to develop fundamentally new toilets. Middle-income countries have little pre-existing wastewater infrastructure and are positioned as the most likely part of the world for practical development of urine source-separating technology.

Urine source separation is a developing research and implementation area. On a global scale, there seem to be few alternatives to recover nutrients and protect surface water against eutrophication. A large portfolio of promising technologies have been developed, but only partial nitrification followed by distillation is close to technical maturity, a technology that can be implemented at the scale of a multi-storage building. Newer chemical and bioelectrochemical technologies hold high promises for smaller-scale applications with the hope of developing small, bathroom-sized applications within the next decade.

References

1. Bengtsson M, Tillman AM (2004) Actors and interpretations in an environmental controversy: the Swedish debate on sewage sludge use in agriculture. Resour Conserv Recycl 42(1):65–82
2. Bischel HN, Özel Duygan BD, Strande L, McArdell CS, Udert KM, Kohn T (2015) Pathogens and pharmaceuticals in source-separated urine in eThekwini, South Africa. Water Res 85:57–65, 11469

3. Boehler MA, Heisele A, Seyfried A, Groemping M, Siegrist H (2015) (NH4)(2)SO4 recovery from liquid side streams. Environ Sci Pollut Res 22(10):7295–7305
4. Borsuk ME, Maurer M, Lienert J, Larsen TA (2008) Charting a path for innovative toilet technology using multicriteria decision analysis. Environ Sci Technol 42(6):1855–1862
5. Bouwman AF, Beusen AHW, Billen G (2009) Human alteration of the global nitrogen and phosphorus soil balances for the period 1970–2050. Glob Biogeochem Cycles 23(4):Gb0a04
6. Bracken P, Wachtler A, Panesar AR, Lange J (2007) The road not taken: how traditional excreta and greywater management may point the way to a sustainable future. Water Sci Technol Water Supply 7:219–227
7. Caby A (2013) Quel intérêt et quelle opportunité de mettre en place une collecte sélective des urines en milieu urbain dense? Etude sur le territoire du SIAAP, Ecole des Ponts ParisTech and AgroParisTech-Engref, Paris
8. Craggs R, Sutherland D, Campbell H (2012) Hectare-scale demonstration of high rate algal ponds for enhanced wastewater treatment and biofuel production. J Appl Phycol 24(3):329–337
9. Czerwionka K, Makinia J (2014) Dissolved and colloidal organic nitrogen removal from wastewater treatment plants effluents and reject waters using physical-chemical processes. Water Sci Technol 70(3):561–568
10. Ekama GA, Wilsenach JA, Chen GH (2011) Saline sewage treatment and source separation of urine for more sustainable urban water management. Water Sci Technol 64(6):1307–1316
11. Fidjeland J, Svensson SE, Vinnerås B (2015) Ammonia sanitization of blackwater for safe use as fertilizer. Water Sci Technol 71(5):795–800
12. Fumasoli A, Etter B, Sterkele B, Morgenroth E, Udert KM (2016) Operating a pilot-scale nitrification/distillation plant for complete nutrient recovery from urine. Water Sci Technol 73(1):215–222
13. Gregory DI, Bumb BL (2006) Factors affecting supply of fertilizer in Sub-Saharan Africa. Discussion paper 24. Agricultural and Rural Development Department, World Bank
14. Hanæus J, Hellström D, Johansson E (1997) A study of a urine separation system in an ecological village in northern Sweden. Water Sci Technol 35:153–160
15. Hellström D, Kärrman E (1997) Exergy analysis and nutrient flows of various sewerage systems. Water Sci Technol 35:135–144
16. Henze M (1997) Waste design for households with respect to water, organics and nutrients. Water Sci Technol 35:113–120
17. Hoglund C, Stenstrom TA, Ashbolt N (2002) Microbial risk assessment of source-separated urine used in agriculture. Waste Manag Res 20(2):150–161
18. Howarth RW, Marino R (2006) Nitrogen as the limiting nutrient for eutrophication in coastal marine ecosystems: evolving views over three decades. Limnol Oceanogr 51(1 II):364–376
19. Jönsson H, Stenström TA, Svensson J, Sundin A (1997) Source separated urine-nutrient and heavy metal content, water saving and faecal contamination. Water Sci Technol 35:145–152
20. Kahiluoto H, Kuisma M, Ketoja E, Salo T, Heikkinen J (2015) Phosphorus in manure and sewage sludge more recyclable than in soluble inorganic fertilizer. Environ Sci Technol 49(4):2115–2122
21. Kuenzle R, Pronk W, Morgenroth E, Larsen TA (2015) An energy-efficient membrane bioreactor for on-site treatment and recovery of wastewater. J Water Sanit Hyg Dev 5(3):448–455
22. Lajaunie-Salla K, Wild-Allen K, Sottolichio A, Thouvenin B, Litrico X, Abril G (2017) Impact of urban effluents on summer hypoxia in the highly turbid Gironde Estuary, applying a 3D model coupling hydrodynamics, sediment transport and biogeochemical processes. J Mar Syst 174:89–105
23. Larsen TA (2015) CO2-neutral wastewater treatment plants or robust, climate-friendly wastewater management? A systems perspective. Water Res 87:513–521
24. Larsen TA, Gujer W (1996) Separate management of anthropogenic nutrient solutions (human urine). Water Sci Technol 34:87–94
25. Larsen TA, Gujer W (1997) The concept of sustainable urban water management. Water Sci Technol 35:3–10

26. Larsen TA, Gebauer H, Gründl H, Künzle R, Lüthi C, Messmer U, Morgenroth E, Niwagaba CB, Ranner B (2015) Blue diversion: a new approach to sanitation in informal settlements. J Water Sanit Hyg Dev 5(1):64–71

27. Ledezma P, Kuntke P, Buisman CJN, Keller J, Freguia S (2015) Source-separated urine opens golden opportunities for microbial electrochemical technologies. Trends Biotechnol 33(4):214–220

28. Lienert J, Larsen TA (2007) Soft paths in wastewater management – the pros and cons of urine source separation. Gaia 16(4):280–288

29. Lienert J, Larsen TA (2010) High acceptance of urine source separation in seven European countries: a review. Environ Sci Technol 44(2):556–566

30. Lienert J, Bürki T, Escher BI (2007) Reducing micropollutants with source control: substance flow analysis of 212 pharmaceuticals in faeces and urine. Water Sci Technol 56(5):87–96

31. Lienert J, Güdel K, Escher BI (2007) Screening method for ecotoxicological hazard assessment of 42 pharmaceuticals considering human metabolism and excretory routes. Environ Sci Technol 41(12):4471–4478

32. Mackey HR, Zheng YS, Tang WT, Dai J, Chen GH (2014) Combined seawater toilet flushing and urine separation for economic phosphorus recovery and nitrogen removal: a laboratory-scale trial. Water Sci Technol 70(6):1065–1073

33. Maurer M, Schwegler P, Larsen TA (2003) Nutrients in urine: energetic aspects of removal and recovery. Water Sci Technol 48:37–46

34. Maurer M, Rothenberger D, Larsen TA (2005) Decentralised wastewater treatment technologies from a national perspective: at what cost are they competitive? Water Sci Technol Water Supply 5:145–154

35. Maurer M, Pronk W, Larsen TA (2006) Treatment processes for source-separated urine. Water Res 40(17):3151–3166

36. McConville JR, Kuenzle R, Messmer U, Udert KM, Larsen TA (2014) Decision support for redesigning wastewater treatment technologies. Environ Sci Technol 48(20):12238–12246

37. McCrackin ML, Cooter EJ, Dennis RL, Harrison JA, Compton JE (2017) Alternative futures of dissolved inorganic nitrogen export from the Mississippi River Basin: influence of crop management, atmospheric deposition, and population growth. Biogeochemistry 133(3):263–277

38. Pahl-Wostl C, Schönborn A, Willi N, Muncke J, Larsen TA (2003) Investigating consumer attitudes towards the new technology of urine separation. Water Sci Technol 48:57–65

39. Rockstrom J, Steffen W, Noone K, Persson A, Chapin FS III, Lambin EF, Lenton TM, Scheffer M, Folke C, Schellnhuber HJ, Nykvist B, de Wit CA, Hughes T, van der Leeuw S, Rodhe H, Sorlin S, Snyder PK, Costanza R, Svedin U, Falkenmark M, Karlberg L, Corell RW, Fabry VJ, Hansen J, Walker B, Liverman D, Richardson K, Crutzen P, Foley JA (2009) A safe operating space for humanity. Nature 461(7263):472–475

40. Roma E, Philp K, Buckley C, Xulu S, Scott D (2013) User perceptions of urine diversion dehydration toilets: experiences from a cross-sectional study in eThekwini municipality. Water SA 39(2):305–311

41. Steffen W, Richardson K, Rockström J, Cornell SE, Fetzer I, Bennett EM, Biggs R, Carpenter SR, De Vries W, De Wit CA, Folke C, Gerten D, Heinke J, Mace GM, Persson LM, Ramanathan V, Reyers B, Sörlin S (2015) Planetary boundaries: guiding human development on a changing planet. Science 347(6223):1259855

42. Sutton PM, Melcer H, Schraa OJ, Togna AP (2011) Treating municipal wastewater with the goal of resource recovery. Water Sci Technol 63(1):25–31

43. Tidåker P, Sjöberg C, Jönsson H (2007) Local recycling of plant nutrients from small-scale wastewater systems to farmland-a Swedish scenario study. Resour Conserv Recycl 49(4):388–405

44. Udert KM, Larsen TA, Gujer W (2006) Fate of major compounds in source-separated urine. Water Sci Technol 54:413–420

45. Udert KM, Buckley CA, Wächter M, McArdell CS, Kohn T, Strande L, Zöllig H, Fumasoli A, Oberson A, Etter B (2015) Technologies for the treatment of source-separated urine in the eThekwini Municipality. Water SA 41(2):212–221

46. Van Drecht G, Bouwman AF, Harrison J, Knoop JM (2009) Global nitrogen and phosphate in urban wastewater for the period 1970 to 2050. Glob Biogeochem Cycles 23(3):Gb0a03

47. Van Puijenbroek PJTM, Bouwman AF, Beusen AHW, Lucas PL (2015) Global implementation of two shared socioeconomic pathways for future sanitation and wastewater flows. Water Sci Technol 71(2):227–233

48. Védry B, Gousailles M, Affholder M, Lefaux A, Bontoux J (2001) From sewage water treatment to wastewater reuse. One century of Paris sewage farms history. Water Sci Technol 43:101–107

49. Vinnerås B, Jönsson H (2013) The Swedish experience with source separation. In: Larsen TA, Udert KM, Lienert J (eds) Source separation and decentralization for wastewater management. IWA, London/New York, pp 415–422

50. Wang H, Gui H, Yang W, Li D, Tan W, Yang M, Barrow CJ (2014) Ammonia nitrogen removal from aqueous solution using functionalized zeolite columns. Desalin Water Treat 52(4–6):753–758

51. Wilsenach JA, van Loosdrecht MCM (2006) Integration of processes to treat wastewater and source-separated urine. J Environ Eng 132(3):331–341

52. Withers PJA, Neal C, Jarvie HP, Doody DG (2014) Agriculture and eutrophication: where do we go from here? Sustainability (Switzerland) 6(9):5853–5875

53. Wood A, Blackhurst M, Hawkins T, Xue X, Ashbolt N, Garland J (2015) Cost-effectiveness of nitrogen mitigation by alternative household wastewater management technologies. J Environ Manag 150:344–354

54. World Bank (2015) Commodity markets outlook. Quarterly outlook, April 2015

55. Zhao Y, Fang Y, Jin Y, Huang J, Bao S, Fu T, He Z, Wang F, Zhao H (2014) Potential of duckweed in the conversion of wastewater nutrients to valuable biomass: a pilot-scale comparison with water hyacinth. Bioresour Technol 163:82–91

56. Jiang F, Chen Y, Mackey HR, Chen GH, Van Loosdrecht MCM (2011) Urine nitrification and sewer discharge to realize in-sewer denitrification to simplify sewage treatment in Hong Kong. Water Science and Technology 64(3):618–626

Tove A. Larsen obtained her master's degree in chemical engineering from the Danish Technical University (DTU) in 1988. At that time, no environmental engineers were educated at DTU, but a flexible study system allowed her to specialize in environmental issues related to water, which was what Dr. Larsen wanted to do. Dr. Larsen got interested in environmental technology when she was in high school and did a project on wastewater treatment plants. At that time, she realized that she could combine her interest in microbiology with the concrete goal of improving environmental quality. Having grown up at the Danish seaside, Dr. Larsen had a very direct connection to the water environment.

Dr. Larsen was offered a Ph.D. position at the Department of Environmental Engineering at DTU, which at that time focused entirely on water. Four years later, Dr. Larsen was the first woman to complete a Ph.D. from this department, where women at that time were usually lab technicians or secretaries.

After her Ph.D., Dr. Larsen became a postdoctoral associate at the Swiss Federal Institute of Science and Technology in Zürich (ETHZ), at the Institute of Environmental Engineering. Dr. Larsen worked at the Chair of Urban Water Management where she taught environmental engineering in the areas of drinking water supply and wastewater treatment. Dr. Larsen was inspired by Gro Harlem Brundtland, the first female Prime Minister of Norway and began to translate the ideas of sustainable development, espoused by Brundtland, to urban water management. The most important part was the idea of source separation as the key to resource-efficient wastewater management. These ideas gained her an offer for a tenure-track position at Eawag, the Swiss Federal Institute of Aquatic Science and Technology, where Dr. Larsen was able to set up and lead a cross-cutting project on urine source separation in wastewater. The project was successful and received a prestigious award for transdisciplinary research from the Swiss Academies of Science in 2008 for "its visionary, innovative and integrative approach to urban water management." Her home institution, Eawag, supported the concept of urine separation and the directorate decided to install urine-separating toilets in all office buildings. Dr. Larsen became tenured at Eawag—as the first woman ever with an engineering degree—and during the early years of her two children, Dr. Larsen could work part time, a welcome fact in Switzerland where working hours are long and fathers at that time were not expected to reduce *their* workload.

In 2011, Eawag received an invitation from the Bill & Melinda Gates Foundation to participate in the Reinvent The Toilet Challenge, and Dr. Larsen decided to set up a concept for a slum toilet based on source separation. Dr. Larsen enjoyed working with colleagues from different disciplines on solving one of the great problems of humanity: sanitation in developing countries. The work resulted in the Blue Diversion Toilet. Despite two Project Innovation Awards from the International Water Association, however, she is still seeking industrial partners without whom the dream of the blue toilet leading to sustainable and dignified sanitation in urban slums will never come true.

In 2014, Dr. Larsen joined the Eawag directorate and in 2017, she became an adjunct professor at DTU-Environment, where Dr. Larsen contributes to a course on urban environmental technologies in developing countries. The number of female faculty at DTU has risen from 0% to 20% since her days as a Ph.D. student, but there is still a long way to go before gender equity in engineering is reached–even in environmental engineering and even in Denmark.

Chapter 7
Environmental Microbiome Analysis and Manipulation

Courtney M. Gardner and Claudia K. Gunsch

Abstract *Bioremediation* is a sustainable environmental treatment technology that harnesses the natural metabolic activities of living organisms to remove contaminants within soil, sediment, and water environments. Bioremediation is generally accepted as being a more cost-effective and sustainable remediation strategy when compared to chemical-based or pump-and-treat systems. Bioremediation treatment strategies are traditionally categorized as either biostimulation or bioaugmentation. *Biostimulation* involves the stimulation of indigenous microorganisms that are capable of degrading contaminants of interest. This treatment approach relies on manipulating site conditions to promote the activity and/or proliferation of microorganisms that are known to metabolize target contaminants of concern in order to overcome rate-limiting metabolic processes. In general, biostimulation involves oxidation-reduction reactions wherein either an electron acceptor (e.g., O_2, Fe^{3+}, or SO_4^{2-}) is added to promote oxidative reduction of a contaminant or an electron donor (e.g., organic substrate) is added to reduce oxidized pollutants.

7.1 Introduction

7.1.1 Bioremediation

Bioremediation is a sustainable environmental treatment technology that harnesses the natural metabolic activities of living organisms to remove contaminants within soil, sediment, and water environments. Bioremediation is generally accepted as being a more cost-effective and sustainable remediation strategy when compared to chemical-based or pump-and-treat systems [1]. Bioremediation treatment strategies are traditionally categorized as either biostimulation or bioaugmentation. *Biostimulation* involves the stimulation of indigenous microorganisms that are

C. M. Gardner · C. K. Gunsch (✉)
Pratt School of Engineering, Department of Civil and Environmental Engineering,
Duke University, Durham, NC, USA
e-mail: ckgunsch@duke.edu

© Springer Nature Switzerland AG 2020
D. J. O'Bannon (ed.), *Women in Water Quality*, Women in Engineering
and Science, https://doi.org/10.1007/978-3-030-17819-2_7

capable of degrading contaminants of interest. This treatment approach relies on manipulating site conditions to promote the activity and/or proliferation of microorganisms that are known to metabolize target contaminants of concern in order to overcome rate-limiting metabolic processes. In general, biostimulation involves oxidation-reduction reactions wherein either an electron acceptor (e.g., O_2, Fe^{3+}, or SO_4^{2-}) is added to promote oxidative reduction of a contaminant or an electron donor (e.g., organic substrate) is added to reduce oxidized pollutants.

Conversely, *bioaugmentation* consists of introducing an exogenous strain or consortium of microorganisms capable of breaking down the contaminant of interest. Bioaugmentation is used either when no indigenous degraders are detected at a contamination site or when indigenous biodegradation is occurring too slowly or inefficiently [2]. When the exogenous strains are able to establish themselves, bioaugmentation is often a more efficient process than biostimulation, occurring three to four times faster in similar environments [3].

A third strategy known as *genetic bioaugmentation* has also emerged. Genetic bioaugmentation is an in situ bioaugmentation strategy that promotes the transfer of genes responsible for contaminant metabolism from exogenous donors to indigenous microbes, thereby increasing overall biodegradation potential [4]. This method relies on high rates of horizontal gene transfer (HGT) of catabolic plasmids that contain the genes involved in biodegradation among phylogenetically similar microorganisms [5] and requires a thorough understanding of the site conditions that are conducive to HGT [6]. However, the survival of the donor exogenous strain is not necessary, so long as the catabolic plasmids are transferred and maintained within the contaminated site [4].

These three strategies have been proven in the past to be effective in multiple aquatic and terrestrial environments, but extreme care and preparation is needed to ensure the successful removal of environmental contaminants [7]. Contaminated sites must be adequately characterized to ensure the survival of exogenous strains used during bioaugmentation. Failure to do so often results in the exogenous microbes being outcompeted by native microorganisms before they are able to effectively degrade target contaminants. In addition, incomplete contaminant transformation may lead to the formation and buildup of more toxic daughter products in the environment [7].

Common bioremediation redox conditions and reactions are shown in Table 7.1. Aerobic biodegradation takes place in the presence of oxygen and relies on the direct microbial oxidation of a contaminant. In aerobic respiration, electrons are generally transferred from the contaminant (electron donor) to oxygen (electron acceptor). This strategy has proven effective for many pollutants, especially non-halogenated organic compounds (e.g., aromatic hydrocarbons and aliphatic hydrocarbons) [8–11]. However, the limiting factor is the availability of oxygen, which must be added to the environment to ensure rapid rates of microbial metabolism and it is often a large cost burden. Anaerobic oxidative bioremediation also takes place in the presence of oxygen but does not use oxygen as the reaction's electron acceptor. Instead, other electron acceptors (e.g., SO_4^{2-}) are used by microorganisms to oxidize the contaminants [12–14].

Table 7.1 Overall redox reactions and contaminant targets for common bioremediation pathways

Process	Redox conditions	Common targets
Aerobic	$O_2 + 4e^- + 4H^+ \rightarrow 2H_2O$	Non-halogenated organics
Anaerobic	Oxidative	Petroleum
	Reductive	Chlorinated solvents, metals
Denitrification	$2NO_3^- + 10e^- + 12H^+ \rightarrow N_2 + 6H_2O$	Varies (NO_3^- as electron acceptor)
Manganese (IV) reduction	$MnO_2 + 2e^- + 4H^+ \rightarrow Mn^{2+} + 2H_2O$	Varies (Mn^{4+} as electron acceptor)
Iron (III) reduction	$Fe(OH)_3 + e^- + 3H^+ \rightarrow Fe^{2+} + 3H_2O$	Varies (Fe^{3+} as electron acceptor)
Sulfate reduction	$SO_4^{2-} + 8e^- + 10H^+ \rightarrow H_2S + 4H_2O$	Varies (SO_4^{2-} as electron acceptor)
Fermentation	$2CH_2O \rightarrow CO_2 + CH_4$	Varies

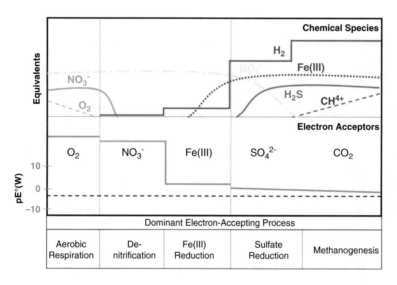

Fig. 7.1 Dominant metabolic process, electron acceptors, and chemical speciation for common microbial bioremediation strategies [3]

Conversely, anaerobic reductive bioremediation takes place in the complete absence of oxygen and requires an input of biologically available organic carbon to proceed [15–17]. Anaerobic reductive bioremediation has also been shown to be effective to treat metal contaminants and reduce their toxicity such as via the reduction of hexavalent chromium to trivalent chromium [18]. However, due to the generally slow growth rates of anaerobic bacteria, anaerobic reductive bioremediation tends to proceed extremely slowly. Each metabolic reaction has a dominant electron acceptor, which is often the acceptor with the highest redox potential.

However, other electron acceptors may be used when this preferred acceptor is depleted (Fig. 7.1). These microbe-based bioremediation strategies may be used for a diverse range of water-contamination scenarios; therefore, understanding the naturally occurring microbial communities in aquatic systems is of particular importance.

7.1.2 Aquatic Microbiomes

7.1.2.1 Microbiomes Associated with Natural Waters

Microorganisms are ubiquitous members of every environmental niche and often serve as drivers of ecological function. Microbial communities play critical roles in all nutrient cycles and are significant drivers in the development and maintenance of aquatic food chains. The collection of these microbial communities and their associated genes, termed *microbiome*, is unique to each environment and its corresponding niche of physical, chemical, and biological attributes [19]. Microbiomes are composed of bacteria, archaea, fungi, protists, and viruses and display a complex range of commensal, synergistic, and pathogenic ecological dynamics, similar to the microbiomes in wastewater treatment. Groundwater and surface water microbiomes are highly variable across pH, temperature, and soil media gradients and are known to contain highly specialized and interdependent members that are unable to survive in laboratory conditions [20]. Aquatic microbiomes are typically dominated by *Bacillus*, *Pseudomonas*, *Rhizobium*, and *Acinetobacter* genera [21]. However, contaminated groundwater microbiomes can be dominated by *Rhodanobacter* sp. due to its ability to adapt to inhospitable environments [22]. Iron-reducing and manganese-reducing bacteria are commonly the dominant degraders in in situ groundwater treatment which utilize Fe^{3+} and Mn^{4+} as electron acceptors, respectively [3, 23].

7.1.2.2 Microbiomes Associated with Water Treatment Systems

Microbiomes are a key player in water treatment systems (e.g., drinking water and wastewater treatment) and are responsible for many of the processes responsible for improving water quality. Drinking water treatment is concerned with improving the quality of raw water sources to produce water that is safe for human consumption. This involves removing sources of chemical and microbial contamination and involves the following steps: crude filtration, coagulation, flocculation, sedimentation, fine filtration, and disinfection. These communities are heavily influenced by the treatment process itself, in addition to physical-chemical factors. The microbiomes associated with drinking water and wastewater treatment are highly diverse and can comprise up to 40 phyla [24]. Filtration and disinfection processes are the largest drivers of drinking water microbiome structure and function. For example, the use of dual media rapid sand filters promotes the proliferation of *Rhizobiales*, *Rhodospirillales*, *Sphingomonadales*, and *Burkholderiales* [24, 25].

Wastewater treatment is the process of converting wastewater (i.e., municipal or industrial waste) into a water effluent that is able to be discharged back to the water cycle or directly reused. Wastewater generated by more than 75% of US households is treated by approximately 16,000 municipal wastewater treatment plants (WWTPs) that are generally broken up into primary treatment, secondary treatment, and tertiary treatment. Secondary treatment is biologically based and relies on microbiomes to degrade suspended and dissolved organic compounds (e.g., biological oxygen demand) [26, 27]. This typically takes the form of activated sludge tanks, rotating biological contactors, aerated lagoon, or constructed wetlands and is performed by aerobic aquatic bacteria. Most of the wasted solids from biological treatment tanks may be further treated via aerobic or anaerobic digestion to produce biosolids. More than 50% of biosolids produced by WWTPs in the USA are used for downstream agriculture or land applications [28]. In general, activated sludge treatment and anaerobic sludge digestion will functionally select for unique microbiome structures. Activated sludge treatment is dominated by *Zooglea*, *Dechloromonas*, *Prosthecobacter*, *Caldilinea*, and *Tricoccus* genera [29]. Anaerobic sludge digestion is dominated by *Bacteroides, Clostridium*, and methanogen genera [30, 31]. The exact microbiome structure varies among individual treatment plants and emerging contaminants such as nanomaterials, pharmaceutically active compounds and flame retardants can affect microbiome structure [32–35], although once established these communities are generally stable over time [36].

7.2 Gene and Contaminant Flow in Water and Water Treatment Systems

7.2.1 Contaminant Transport in Water

There are currently more than 60,000 chemicals registered under the EPA Toxic Substances Control Act (TSCA), and an increasing number of these compounds show evidence of negatively impacting human or ecological health [37] and have been detected in natural waters and water treatment systems (Table 7.2). Groundwater and surface water contamination is of significant concern for both humans and the environment. Groundwater contamination is nearly always the result of anthropogenic activities and is particularly prevalent in geographic area with high population densities. In general, contaminant transport is dictated by a number of factors including chemical solubility, mobility, and reactivity. Contaminants that are highly soluble such as salts and some metals are readily able to partition from surface soils to groundwater. Other compounds that display higher hydrophobicity are more likely to initially partition out of surface waters and adsorb to other hydrophobic surfaces such as soils and sediments. This adsorption protects contaminants from chemical or microbial degradation and allows them to have significantly longer residence times in these environments. However, during precipitation and runoff events, small fractions of these hydrophobic compounds may desorb from soil surfaces and

Table 7.2 Emerging and historical contaminants in groundwater, drinking water, and wastewater [38]

Compound	CAS	Structure/class	Common use
1,4-Dioxane	123-91-1	Ether	Solvent
Acrolein	107-02-8	Aldehyde	Aquatic herbicide
Antibiotics	Varies	Varies	Treatment of bacterial infections
Antibiotic resistance genes (e.g., *tet*)	N/A	DNA-based	Confer bacterial resistance to antimicrobials and heavy metals
Atrazine	1912-24-9	Triazine	Herbicide
Cyanotoxins	N/A	N/A	Naturally released by Cyanobacteria
Estradiol	50-28-2	Steroid	Pharmaceutical
Gemfibrozil	25812-30-0		Pharmaceutical
Perfluorooctanoic acid (PFOA)	335-67-1	Fluorinated carboxylic acid	Water and oil repellent
Perfluorooctanesulfonic acid (PFOS)	1763-23-1	Fluorinated carboxylic acid	Water and oil repellent
Permethrin	52645-53-1	Pyrethroid	Insecticide
Metals/nanoparticles (e.g., Ag, Cu)	Varies	Varies	Varies

migrate into the water table. Microbial contamination (e.g., pathogens) in groundwater generally originates from effluents from septic tanks and other cesspools.

7.2.2 Emerging Contaminants: Environmental Antibiotic Resistance

As noted in Table 7.2, a wide chemical range of anthropogenic contaminants have been detected in natural waters and water treatment systems. Microbial contamination of drinking water has historically been primarily focused on removing microbial pathogens such as Escherichia coli or Naegleria fowleri. However, in recent years, environmental transport of antibiotic resistance genes (ARGs) and metals has raised significant concerns among scientists due to rising rates of antibiotic- and multidrug-resistant (MDR) bacteria.

Antibiotic resistance rates have increased dramatically in both clinical and environmental bacterial populations over the past several decades [39–43]. These sources are believed to increase antibiotic resistance by providing a significant and consistent selective pressure (i.e., antibiotics or heavy metals) which may promote antibiotic-resistant bacteria maintenance and the propagation of ARGs [44]. However, an upsurge in the number of bacteria that have acquired immunity to a single antibiotic as well as MDR bacteria has been noted in a wide array of environments, due in part to the widespread and increased use of antibiotics [44–46]. Resistance genes for tetracyclines and β-lactamases and several other resistance genes have been shown to have some of the highest rates of increase among soil bacteria [47, 48].

7.2.2.1   Antibiotics, ARGs, and Metals in Natural Water Systems

ARGs and metals have been documented in a wide array of natural water environments, including groundwater, surface water, drinking water treatment plant influent and effluent, and wastewater treatment plant influent and effluent. Common metal contaminants of concern include arsenic, cadmium, copper, nickel, lead, silver, and mercury. Many of these metals can enter natural water systems through point sources such as acid mine drainage sites and landfills as well as improper disposal of metal-containing consumer products or industrial waste. In addition to having negative impacts on ecological health, metal contamination in natural and engineered water systems is a significant concern because of their relationship with environmental antibiotic resistance [32]. Many ARGs are located on bacterial plasmids that also contain heavy metal-resistant genes; therefore, the widespread land applicaiton of these compounds may lead to the selection of antibiotic- and multidrug-resistant bacteria in agricultural environments.

7.2.2.2   Proliferation of ARGs and Metal Water Treatment

Microbiome exposure to ARGs and metals is not limited to natural water environments. A wide array of ARGs have been documented in municipal wastewater, hospital wastewater, dairy and swine farm lagoons, groundwater, and surface water [49–52]. More recently, ARGs have been detected at significant levels in extracellular deoxyribonucleic acid (DNA) within river waters and sediments [53]. This suggests that these genetic elements are much more mobile than initially thought and have the potential to survive outside of bacterial cells and affect microbial communities in several disparate microbiomes. DNA is able to persist in both drinking water and wastewater effluents and can be difficult to remove, posing a significant public and ecological health concern [54–56].

Municipal WWTPs have been shown to contain both antibiotic-resistant bacteria and antibiotic chemicals. In fact, WWTPs are widely thought to be a significant contributor to the dissemination of ARGs in the environment. ARGs detected in WWTP environments include *sulI, sulII, erm(B), erm(F), bla, nptII, tet(O), tet(W), tet(C), tet(G)*, and *tet(X)* among others [57–60] (see Table 7.3).

Any present ARGs that remain in digester tanks may also be transferred to biosolids—50–60% of which is commonly used as agricultural fertilizer annually. The relatively high background levels of antibiotic contaminants in wastewaters suggest that there is a further possibility that free ARGs may be horizontally transferred to WWTP bacteria, thus further increasing antibiotic resistance in the environment. The ability of WWTP bacteria to take up these transgenes in this environment is a factor of (1) the observed rates of HGT in anaerobic environments, (2) the abundance of genes of interest within WWTP anaerobic digesters, and (3) the presence of a selective pressure.

The sources of the ARGs and metals and methods of dissemination are difficult to track due to the sheer number of potential sources and the complexity of inputs feeding into water treatment system. In drinking water and wastewater treatment

Table 7.3 Percentage of antibiotic-resistant bacteria detected in natural waters and water treatment systems

Antibiotic	Hospital effluent (%)	Activated sludge (municipal WWTP) (%)	Effluent (municipal WWTP) (%)	Surface water (%)	Drinking water (%)
Vancomycin	6.8 (±5.0)	11 (±3.8)	15 (±10)	2.3 (±0.5)	20 (±10)
Ceftazidime	45 (±21)	44 (±17)	27 (±17)	11 (±1.6)	5.1 (±2.4)
Cefazolin	58 (±23)	39 (±16)	39 (±20)	8.1 (±0)	48 (±27)
Penicillin G	71 (±25)	30 (±8.0)	20 (±6.7)	31 (±3.3)	43 (±26)
Imipenem	8.1 (±3.5)	2.8 (±0.2)	0.6 (±0.4)	0.4 (±0.1)	0

Adapted from Schwartz et al. [58]

systems, most ARGs enter through influents in the form of intracellular DNA contained within bacteria. However, a smaller fraction of ARGs have also been documented in extracellular DNA that exists outside of bacterial cells, likely released as a result of cell lysis. This free DNA is able to persist in environmental systems when bound to positively charged colloid particles (e.g., clays) and is highly resistant to both enzymatic and physicochemical degradation [61, 62]. More than 90% of ARGs that enter WWTPs in influent streams is removed through settling during secondary treatment (e.g., activated sludge digestion) and ends up in anaerobic sludge digester tanks [53]. While high operation temperature (>45 °C) and solid retention times (SRTs) are associated with the removal of some of these ARGs, the bulk of them remain after anaerobic digestion—more than 10^{10} gene copies per gram of anaerobic sludge for some ARGs [63]. Increased persistence may be due to the protection offered by DNA-clay complexes. ARGs that do not settle out either partition into anaerobic digesters or remain in waters bound to be discharged into surface water. Though some effluents undergo tertiary treatment before leaving the WWTP, only some methods (e.g., UV and chlorination) are effective for removing DNA.

7.2.3 Environmental Contaminants as Drivers of Microbiome Evolution

In general, it has been noted that HGT is a major driver of evolution within environmental microbiomes. In heavy metal-contaminated groundwater systems, genes responsible for heavy metal resistance, such as Co^{2+}, Zn^{2+}, and Cd^{2+}, are under large selection pressures and are transferred or duplicated within these microbiomes at exceedingly high rates [22]. However, even a small concentration of these metals as nanoparticles in wastewater-generated biosolids may have observable impacts on ecosystem functions [54]. As the number of emerging contaminants detected in natural waters and water treatment systems increases, so, too, does the need for technologies that are able to mitigate these public health threats. Thus, it is of paramount importance to understand how these microbiomes interact with and are impacted by emerging contaminants of concern.

7.3 Modeling Relationships Between Engineering Design and Microbiomes/Water Quality

7.3.1 Strategies to Characterize and Annotate Environmental Microbiomes

Deciphering relationships among communities is difficult for environmental microbiomes. An overwhelming number of niches and environmental factors must be taken into account as well as the comparative degree of heterogeneity to understand these systems. It is currently estimated that only a small fraction of bacterial species is known. Large knowledge gaps also exist in describing other microorganisms such as fungi and archaea as well as their corresponding metabolic functions in the environment. For example, less than 2% of data generated by non-targeted metabolomics studies—which characterize the metabolite compounds present in a cell—are able to be matched to known chemical compounds, with few mapped to known biochemical pathways [20]. Furthermore, more than half of all known microbial phyla lack a single fully characterized species. Thus, additional work and innovation is needed to understand these systems and the roles of individual microorganisms within them. Many different approaches are needed to characterize and annotate the structures and functions of aquatic microbiomes. The most important questions currently facing this area of research include:

1. How do variations in the environmental conditions of natural waters and water treatment systems impact the development of their corresponding microbiomes?
2. Which ecological functions can be predicted by characterizing aquatic microbiome structures?
3. In what ways are chemical and genetic contaminants impacting microbiome structure and functioning in natural and engineered aquatic systems?

The challenges posed by each of these questions are formidable, given the enormous diversity and number of the world's microbiomes and their corresponding functions. However, substantial advancements have been made with regard to the technologies available to investigate these systems. Current efforts are focusing on understanding the relationships among environmental conditions and microbiome formation and ecological function and are being spearheaded by the Unified Microbiome Initiative. The Unified Microbiome Initiative was announced in 2015 with the goal of developing innovative molecular biotechnology tools to accelerate basic characterization of environmental microbiomes as well as facilitating these findings to other applications. This mission has the ability to augment the tools available to delineate the impacts of aquatic microbiome structure and function on environmental quality and has an incredible potential to drive innovations in water treatment technologies.

7.3.2   Tools to Characterize Microbiome Structure and Function

7.3.2.1   DNA Sequencing

One of the most promising tools available to study the structure of environmental microbiomes is DNA-based next-generation sequencing (NGS). One of the first DNA sequencing methods was developed in 1977 by Frederick Sanger, termed *Sanger sequencing*, and relies on chain termination to determine the DNA template nucleotide sequence. While ideal for relatively short DNA templates, approximately 700–900 base pairs, this method is cost prohibitive when investigating large sequences or entire microbiomes. These factors have significantly limited the use of Sanger sequencing for characterizing environmental microbiomes and ultimately led to the development of high-throughput NGS in the 1990s.

Historically, the most widely used sequencing platforms include Ion Torrent semiconductor sequencing, 454 pyrosequencing, Illumina sequencing, and Nanopore sequencing (Table 7.4). Illumina Miseq and Hiseq sequencing systems currently dominate studies of human, terrestrial, and aquatic microbiomes due largely to its high accuracy rates, number of reads generated per run, and relative low cost. Illumina excels in the characterization of both amplicon-based and full metagenomic DNA libraries. Amplicon-based metagenomic libraries require the amplification of target DNA fragments (e.g., *16S* for bacterial and archaea communities or *ITS/18S* for fungal communities) using end-point polymerase chain reaction (PCR). This is particularly useful for characterizing the microbiome community structures associated with individual community subtypes but gives no insight into community dynamics or functions. Full metagenomic sequencing is performed at the Unified Microbiome Initiative on the Illumina Hiseq platform and allows for the complete sequencing of all DNA

Table 7.4 Comparison of high-throughput NGS methods [64]

Method	Read length (bp)	Accuracy (%)	Reads per run	Time per run	Cost per million bases
Ion torrent sequencing	<600	98	<80 million	2 hours	$1.00
Pyrosequencing (454)	700	99.9	1 million	24 hours	$10
Illumina sequencing	MiSeq: 50–600 bp; HiSeq 2500: 50–500 bp; HiSeq 3/4000: 50–300 bp	99.9	MiSeq: 1–25 Million; HiSeq 2500: 300 million–2 billion, HiSeq 3/4000 2.5 billion	1–11 days	$0.05–0.15
Nanopore sequencing	Up to 500 kb	92–97	Varies by read length	1 minute to 48 hours	Varies by read length
Sanger sequencing	400–900	99.9	N/A	20 minutes to 3 hours	$2400

extracted from an environmental sample. This method gives insight not only into the microbial community structures in a particular niche, but also the complete genomic content of each sequenced microbe.

7.3.2.2   RNA Sequencing

Large amounts of data can be obtained from DNA-based sequencing studies of aquatic microbiome structures, but the interpretation of that data is quite limited. While the detection of microbial DNA is suggestive of its presence within a given environment, it does not provide evidence that a microbe is transcriptionally active or even alive. This extracellular DNA can be incorporated into NGS DNA libraries and sequenced as if it were associated with a living cell. Predictive modeling programs do exist (e.g., PICRUSt), but these programs rely on phylogeny and marker gene data that are only suggestive of metabolic activity. Transcriptomic sequencing, using RNA templates, is able to address this issue when analyzing environmental microbiomes. Transcriptomic sequencing refers to the set of all RNA molecules (i.e., mRNA, tRNA, and rRNA) within a single-cell or microbiome community. Unlike genomic content, which is largely fixed within a cell or population, the transcriptome may vary widely across niches and environmental conditions as genes are up- or downregulated. Transcriptomic analyses may be performed with various techniques such as DNA and RNA microarrays and RNA-Seq, another NGS platform.

7.3.2.3   Metabolomic and Proteomic Analysis

Metabolomic analysis takes transcriptomic assessments one step further by annotating the biochemical pathways and metabolites generated by both single cells and complex microbiome communities. The *metabolome*, therefore, is the summation of these biochemical pathways and metabolites within a cell, tissue, organism, or environment and includes metabolic intermediates, signaling molecules, and secondary metabolites. Metabolomes are highly specific to both microbiome composition and environmental parameters. Metabolomes are characterized by first separating aggregates of analytes (e.g., via gas chromatography, high-performance liquid chromatography, or capillary electrophoresis) before detecting individual compounds (e.g., via mass spectrometry or secondary ion mass spectrometry). This type of analysis can be particularly useful when trying to determine the biochemical pathways associated with a biostimulation or bioaugmentation treatment strategy, as it can provide direct information about the metabolic pathways involved in contaminant degradation as well as potential daughter products that may be produced in a bioremediation scenario. Metabolomics also provides an instantaneous snapshot of the physiology and metabolism of a single cell and improves current understanding of how individual cells function in complex mixed communities.

Proteomics is concerned with discovering and annotating the proteins produced by individual cells as well as collections of cells. Similar to transcriptomics, the study of an environmental proteome is made more complicated by the fact that a proteomic profile varies between individual cells and may change drastically across varying environmental conditions. This makes it difficult to obtain a "core proteome" for any given environment. Proteomes in environmental samples may be detected using immunoassays, mass spectrometry, or protein chips [65]. While generally more expensive than immunoassays, MS analysis allows for the high-throughput analysis of large-scale environmental studies [66].

Each of these platforms generates incredible volumes of data—up to several gigabytes for each sample. Analyzing the simpler data sets produced by amplicon-based Illumina Miseq sequencing can take weeks or months for a single data set. This is further complicated by the large amounts of metadata that typically accompany environmental studies, which must also be analyzed and correlated with microbiome data. As a result, bioinformatics tools have lagged behind in their ability to keep up with the volume and nature of NGS data [67–69]. Currently, there is a critical need to develop more bioinformatics tools to facilitate the analysis and integration of -omics data with environmental metadata.

7.3.3   Assessing Contaminant Impact on Microbiome Structure and Function

One of the most promising applications of these bioinformatics technologies is the study of how aquatic microbiomes are impacted by acute and chronic exposures to environmental contaminants in natural and engineered water environments. These potential impacts can be measured in two ways: alterations to microbiome community structure (i.e., shifts in diversity or abundance of observed microbes) or changes in microbiome function (i.e., the gain or loss of biological genes and processes). Microbiome structures can be elucidated with targeted amplicon-based or shotgun metagenomic sequencing. Inferences can be drawn concerning microbes that become less abundant—or disappear altogether—after exposure to an environmental contaminant and their likely susceptibility to the contaminant of interest. The remaining microbes likely display resistance to the contaminant's toxicity or may be able to actively metabolize it. These taxa may be investigated further to determine their potential as candidate bioremediation targets. However, impacts on microbiome structure do not necessarily imply alterations in microbiome function or ecological health. These parameters can be assessed using transcriptomic, metabolomic, or proteomic approaches. Many microbiomes have been found to display a high degree of functional redundancy. Functional redundancy refers to biochemical or metabolic processes that are undertaken by more than one microbe inhabiting the same environment. In wastewater treatment, redundancy is found in BOD removal, an essential bioprocess that is performed by many aquatic microbes. However, there are some equally essential for water quality management that do not have a high degree of functional redundancy. One such process is nitrogen removal, which is performed almost exclusively by nitrogen-cycling bacteria. These bacteria are also highly sensitive to changes in their environment and have been observed to be significantly negatively impacted by the presence of many environmental contaminants in drinking water and wastewater treatment systems [70, 71]. Thus, it will be essential to determine how common exposures to environmental contaminants affect microbiome community structures and processes that are integral to improving water quality.

7.4 Machine Learning to Identify Trends Among Water Microbiomes

7.4.1 Machine Learning

One of the considerable challenges facing the study of environmental microbiomes is the assimilation of genomic, transcriptomic, metabolomic, and proteomic information of a given system in order to understand the overall structure and function of microbial communities as well as the role of individual species within those communities. The burgeoning field of machine learning offers a potential solution to this problem. Machine learning is the application of artificial intelligence that focuses on the development of computer programs that access, analyze, and identify trends within large data sets without being explicitly programmed. This is distinct from traditional computing software algorithms; these "expert systems" are knowledge-based systems that are pre-governed by rules and constraints on a given topic. The strengths of these systems are derived from their collection and implementation of both factual and heuristic knowledge.

Conversely, machine learning systems are not governed by a set of operator-defined rules and instead "learn" rules from the data sets themselves [72]. These systems work by identifying patterns within highly complex data and metadata sets and are applied to computing tasks in which the use of explicit expert systems would be impractical. The types of algorithms that are utilized for machine learning are diverse and include artificial neural networks, Bayesian networks, and genetic algorithms [73]. These algorithms can be employed for a wide variety of applications, from email filtering to disease detection and prognosis [74]. The ultimate goal of machine learning models for environmental microbiomes is the prediction of structure, transcripts, and ecological function from limited data sets across multiple scales—from individual cells to entire niches [75].

7.4.1.1 Limitations

Machine learning offers the prospect of unparalleled insights into environmental microbiomes, although it is not without its limitations. Machine learning requires a large quantity of input data—potentially millions of data points for more complex environmental studies—to reach acceptable performance levels, which is often infeasible [75]. The variety of input data is another limiting factor [76]. This is particularly problematic for studies investigating environmental microbiomes due to the heterogeneity of a given matrix [77].

Aquatic microbiomes can vary greatly based on small variations in sampling depth, water pH, or nutrient composition. These biases are compounded in molecular biotechnology, which introduces additional biases within nucleic acid isolation protocols and PCR amplification of template nucleic acids. Increased standardization and transparency in the sampling and processing of environmental microbiomes must therefore be a priority. Current efforts by groups such as *The Earth Microbiome Project* are working to address these issues by publishing explicit methods for the collection of samples and metadata, DNA isolation, prokaryotic and fungal DNA library preparation, and bioinformatics analysis [76, 78]. Machine learning has the potential to revolutionize the way we are able to study and address issues facing global water quality, particularly with regard to understanding how to best engineer water treatment systems to address water contamination.

7.5   Applications in Water and Wastewater Treatment

There are many water sources and treatment systems that scientists and engineers can manipulate using prebiotics, probiotics, and vectors to improve water quality and sustainability. The availability and diversity of powerful biotechnology tools (e.g., NGS, transcriptomics, and machine learning) will enable the elucidation of trends in aquatic contaminant transport and biodegradation that can be exploited for precision bioremediation in both natural and engineered water systems. Precision bioremediation is the combined use of molecular biotechnology, bioinformatics, and environmental engineering principles for environmental contaminant removal and offers a new way to adapt treatment technologies that ensure environmental quality. Precision bioremediation is adaptable across multiple scales of contamination and microbiome manipulation. It is predicated upon a thorough understanding of site-specific microbiomes as well as physicochemical parameters that are interacting with both the microbiomes and the contaminants themselves. One of the most effective ways of employing precision bioremediation in the field is through in situ microbiome engineering.

7.5.1   Frameworks for Microbiome Engineering

Precision bioremediation using in situ microbiome engineering techniques is a promising future direction in environmental quality management and innovation and can be implemented at multiple scales using the design-build-test paradigm. This framework aims to determine the ideal bioremediation strategy based on the

Table 7.5 Potential in situ methods for microbiome engineering in water treatment [79]

Method	Common examples	Predominant targets	Mechanism of action
Prebiotics	Dietary fibers (inulin), polysaccharides (oligosaccharides)	Variable	Promote targeted bacterial growth
Probiotics	Firmicutes (Lactobacillus), Actinobacteria (Bifidobacteria), Proteobacteria	Variable	Compete for nutrients, produce antimicrobials, modulate environment
Probiotic consortium	Endophytic phytoaugmentation (soils or constructed wetlands)	Environmental contaminant degradation	Establish native rhizosphere community
Engineered probiotics	Streptococcus	Human pathogens (e.g., MDR bacteria)	Inactivation via CRISPR-Cas9
Bacteriophage	Specific phage strains or consortia	Variable	Cell lysis (lytic), genomic integration (lysogenic), transduction
Genetic bioaugmentation	Pseudomonas putida	Environmental contaminant degradation	Transfer of catabolic genes and plasmids

unique contaminants of interest, physicochemical properties of the contaminated environment, and endogenous microbiomes. Much of this information can be collected using -omics-based approaches and assimilated using machine learning principles to develop a highly specialized and effective strategy to manipulate aquatic microbiome processes and improve water quality.

Microbiome manipulations fall under three main categories: prebiotic interventions, probiotic interventions, and genomic interventions under this paradigm (Table 7.5). Prebiotic interventions involve the addition of a beneficial chemical compound that is able to promote the growth of targeted beneficial microbes or enhance specific metabolic activities in a given environment. Unfortunately, it is often difficult to predict the specificity of their interactions with environmental microbiomes, thus making it possible for undesirable microbes to also proliferate (e.g., bacterial pathogens). However, recent advances in high-throughput NGS, transcriptomics, and metabolomics have made it possible to begin addressing this shortcoming. Functional metagenomics and transcriptomics can provide a more nuanced view of aquatic microbiome constituents and community dynamics, which facilitates the identification of prebiotic compounds specific to target microbes. Proteomic analysis can provide insights into characterizing the enzymes responsible for contaminant degradation (e.g., monooxygenases), which could be used to develop biologically based remediation products to deploy during in situ bioremediation. For example, the addition of prebiotic oxidizing compounds could be directly applied to waters and soils to facilitate the removal of environmental contaminants and extracellular ARGs.

Probiotic-based microbiome engineering is similar to traditional bioaugmentation in that it relies on the introduction of an exogenous microbe strain or consortium in order to alter the function of an environmental microbiome. Precision bioremediation could take this one step further by assessing the structure and community dynamics within these environments using functional metagenomics and transcriptomics. In addition to providing a more thorough understanding of natural biodegradation potential, it would also provide a more refined lens with which to view potential manipulation strategies to ensure the survival of the exogenous strains. One example of ex situ microbiome engineering involves the construction of synthetic communities that recreate the core biochemical and ecological processes of a natural system but contain defined members that can be easily manipulated for various endpoints. Contaminated water could be pumped from the contamination site to a holding tank containing synthetic microbial communities capable of biodegradation. The selection of hydraulic residence times and microbial consortia would be informed by the relevant contaminants of interest.

DNA-based interventions may be used for in situ aquatic microbiome engineering, in addition to the application of chemical and cellular modulators. These interventions take place on a genomic scale and introduce new genes to a microbial community through transduction, translation, or conjugation. For example, transduction via lysogenic bacteriophages present a unique opportunity for aquatic microbiome engineering [80, 81]. Phages may be highly specific or polyvalent, able to target multiple bacterial species at a time. As such, they may easily be used across multiple scales. Genetic bioaugmentation can be used to a similar effect to introduce novel metabolic or biochemical processes within an indigenous community. As with the introduction of phages, this method does not require the survival of the donor organism [82]. Genome-editing technologies such as CRISPR-Cas9 also have the potential to edit the functions of aquatic microbiomes to promote the removal of traditional environmental contaminants. However, unlike phages and genetic bioaugmentation, CRISPR may prove equally useful in combatting the transport and proliferation of non-traditional and unique emerging contaminants such as ARGs. CRISPR could be used in ARG "hotspots" such as wastewater and drinking water treatment plants to remove or turn off intracellular ARGs.

In conclusion, ensuring the elimination of toxic environmental contaminants from natural waters and water treatments is of paramount importance for maintaining and improving water quality and ecological sustainability. The advancement of molecular biotechnology is enabling the development of new water treatment technologies and methodologies. Aquatic microbiome-based bioremediation strategies which combine-omics analysis and machine learning are one of the most promising approaches due to their sustainability and provide substantial opportunities for research exploration. Successful adoption and implementation of these strategies could make significant strides in improving water quality and ensuring global human health.

References

1. Gavrilescu M (2009) Emerging processes for soil and groundwater cleanup-potential benefits and risks. Environ Eng Manag J 8(5):1293–1307
2. Jianlong W, Xiangchun Q, Libo W, Yi Q, Hegemann W (2002) Bioaugmentation as a tool to enhance the removal of refractory compound in coke plant wastewater. Process Biochem 38(5):777–781
3. EPA (2013). Introduction to in situ bioremdiation of groundwater. Office of Solid Wastew and Emergency Response, EPA/542/R-13-018
4. Ikuma K, Gunsch CK (2012) Genetic bioaugmentation as an effective method for in situ bioremediation: functionality of catabolic plasmids following conjugal transfers. Bioengineered 3(4):236–241
5. Ikuma K (2011) The effect of select biological and environmental factors on the horizontal gene transfer and functionality of the TOL plasmid: a case study for genetic bioaugmentation
6. Ikuma K, Gunsch CK (2013) Functionality of the TOL plasmid under varying environmental conditions following conjugal transfer. Appl Microbiol Biotechnol 97(1):395–408
7. Vogel TM, McCARTY PL (1985) Biotransformation of tetrachloroethylene to trichloroethylene, dichloroethylene, vinyl chloride, and carbon dioxide under methanogenic conditions. Appl Environ Microbiol 49(5):1080–1083
8. Chae S-R, Hunt DE, Ikuma K, Yang S, Cho J, Gunsch CK, Liu J, Wiesner MR (2014) Aging of fullerene C 60 nanoparticle suspensions in the presence of microbes. Water Res 65:282–289
9. Fritsche W, Hofrichter M (2008) Aerobic degradation by microorganisms. In: Biotechnology set, 2nd edn, pp 144–167
10. Gunsch CK, Kinney KA, Szaniszlo PJ, Whitman CP (2006) Quantification of homogentisate-1, 2-dioxygenase expression in a fungus degrading ethylbenzene. J Microbiol Methods 67(2):257–265
11. Verce MF, Gunsch CK, Danko AS, Freedman DL (2002) Cometabolism of cis-1, 2-dichloroethene by aerobic cultures grown on vinyl chloride as the primary substrate. Environ Sci Technol 36(10):2171–2177
12. Bossert I, Young L (1986) Anaerobic oxidation of p-cresol by a denitrifying bacterium. Appl Environ Microbiol 52(5):1117–1122
13. Gibson D, Koch J, Kallio R (1968) Oxidative degradation of aromatic hydrocarbons by microorganisms. I. Enzymic formation of catechol from benzene. Biochemistry 7(7):2653–2662
14. Rueter P, Rabus R, Wilkest H, Aeckersberg F, Rainey FA, Jannasch HW, Widdel F (1994) Anaerobic oxidation of hydrocarbons in crude oil by new types of sulphate-reducing bacteria. Nature 372(6505):455
15. Bouwer EJ, Rittmann BE, McCarty PL (1981) Anaerobic degradation of halogenated 1-and 2-carbon organic compounds. Environ Sci Technol 15(5):596–599
16. Leahy JG, Colwell RR (1990) Microbial degradation of hydrocarbons in the environment. Microbiol Rev 54(3):305–315
17. Middeldorp PJ, Luijten ML, Pas BAvd, Eekert MHv, Kengen SW, Schraa G, Stams AJ (1999) Anaerobic microbial reductive dehalogenation of chlorinated ethenes. Biorem J 3(3):151–169
18. Wang P-C, Mori T, Komori K, Sasatsu M, Toda K, Ohtake H (1989) Isolation and characterization of an *Enterobacter cloacae* strain that reduces hexavalent chromium under anaerobic conditions. Appl Environ Microbiol 55(7):1665–1669
19. Hug LA, Baker BJ, Anantharaman K, Brown CT, Probst AJ, Castelle CJ, Butterfield CN, Hernsdorf AW, Amano Y, Ise K (2016) A new view of the tree of life. Nat Microbiol 1:16048
20. Comolli LR, Banfield JF (2014) Inter-species interconnections in acid mine drainage microbial communities. Front Microbiol 5:367
21. Martin MS, Santos IC, Carlton DD Jr, Stigler-Granados P, Hildenbrand ZL, Schug KA (2018) Characterization of bacterial diversity in contaminated groundwater using matrix-

assisted laser desorption/ionization time-of-flight mass spectrometry. Sci Total Environ 622–623:1562–1571

22. Hemme CL, Green SJ, Rishishwar L, Prakash O, Pettenato A, Chakraborty R, Deutschbauer AM, Van Nostrand JD, Wu L, He Z (2016) Lateral gene transfer in a heavy metal-contaminated-groundwater microbial community. mBio 7(2):e02234-15

23. Anderson RT, Lovley DR (1997) Ecology and biogeochemistry of in situ groundwater biore-mediation. Adv Microb Ecol 15:289–350

24. Proctor CR, Hammes F (2015) Drinking water microbiology—from measurement to manage-ment. Curr Opin Biotechnol 33:87–94

25. Pinto AJ, Xi C, Raskin L (2012) Bacterial community structure in the drinking water microbi-ome is governed by filtration processes. Environ Sci Technol 46(16):8851–8859

26. Kinney CA, Furlong ET, Zaugg SD, Burkhardt MR, Werner SL, Cahill JD, Jorgensen GR (2006) Survey of organic wastewater contaminants in biosolids destined for land application. Environ Sci Technol 40(23):7207–7215

27. Wu C, Spongberg AL, Witter JD, Fang M, Czajkowski KP (2010) Uptake of pharmaceutical and personal care products by soybean plants from soils applied with biosolids and irrigated with contaminated water. Environ Sci Technol 44(16):6157–6161

28. Hale RC, La Guardia MJ, Harvey EP, Gaylor MO, Mainor TM, Duff WH (2001) Flame retar-dants: persistent pollutants in land-applied sludges. Nature 412(6843):140–141

29. Zhang T, Shao M-F, Ye L (2012) 454 pyrosequencing reveals bacterial diversity of activated sludge from 14 sewage treatment plants. ISME J 6(6):1137

30. Nelson MC, Morrison M, Yu Z (2011) A meta-analysis of the microbial diversity observed in anaerobic digesters. Bioresour Technol 102(4):3730–3739

31. Li B, Yang Y, Ma L, Ju F, Guo F, Tiedje JM, Zhang T (2015) Metagenomic and network analy-sis reveal wide distribution and co-occurrence of environmental antibiotic resistance genes. ISME J 9(11):2490

32. Alito CL, Gunsch CK (2014) Assessing the effects of silver nanoparticles on biological nutri-ent removal in bench-scale activated sludge sequencing batch reactors. Environ Sci Technol 48(2):970–976

33. Gwin CA, Lefevre E, Alito CL, Gunsch CK (2018) Microbial community response to silver nanoparticles and Ag+ in nitrifying activated sludge revealed by ion semiconductor sequenc-ing. Sci Total Environ 616:1014–1021

34. Wang S, Gunsch CK (2011) Effects of selected pharmaceutically active compounds on treat-ment performance in sequencing batch reactors mimicking wastewater treatment plants opera-tions. Water Res 45(11):3398–3406

35. Lefevre E, Cooper E, Stapleton HM, Gunsch CK (2016) Characterization and adaptation of anaerobic sludge microbial communities exposed to tetrabromobisphenol A. PLoS One 11(7):e0157622

36. Munck C, Albertsen M, Telke A, Ellabaan M, Nielsen PH, Sommer MO (2015) Limited dis-semination of the wastewater treatment plant core resistome. Nat Commun 6:8452

37. Worth A, Balls M (2002) Alternative (non-animal) methods for chemicals testing: current sta-tus and future prospects a report prepared by ECVAM and the ECVAM Working Group on Chemicals. ATLA-NOTTINGHAM- 30:1–3

38. EPA (2008), Contaminant candidate list (CCL) and regulatory determination (CCL4), https://www.epa.gov/ccl/chemical-contaminants-ccl-4

39. Gerhard WA, Choi WS, Houck KM, Stewart JR (2017) Water quality at points-of-use in the Galapagos Islands. Int J Hyg Environ Health 220(2):485–493

40. Goossens H, Ferech M, Vander Stichele R, Elseviers M, E.P. Group (2005) Outpatient anti-biotic use in Europe and association with resistance: a cross-national database study. Lancet 365(9459):579–587

41. Khachatourians GG (1998) Agricultural use of antibiotics and the evolution and transfer of antibiotic-resistant bacteria. Can Med Assoc J 159(9):1129–1136

42. Neuhauser MM, Weinstein RA, Rydman R, Danziger LH, Karam G, Quinn JP (2003) Antibiotic resistance among gram-negative bacilli in US intensive care units: implications for fluoroquinolone use. JAMA 289(7):885–888

43. Witte W (1998) Medical consequences of antibiotic use in agriculture. Science 279(5353): 996–997

44. Allen HK, Donato J, Wang HH, Cloud-Hansen KA, Davies J, Handelsman J (2010) Call of the wild: antibiotic resistance genes in natural environments. Nat Rev Microbiol 8(4):251–259

45. Knapp CW, Dolfing J, Ehlert PA, Graham DW (2009) Evidence of increasing antibiotic resistance gene abundances in archived soils since 1940. Environ Sci Technol 44(2):580–587

46. Pruden A, Pei R, Storteboom H, Carlson KH (2006) Antibiotic resistance genes as emerging contaminants: studies in northern Colorado. Environ Sci Technol 40(23):7445–7450

47. Chopra I, Roberts M (2001) Tetracycline antibiotics: mode of action, applications, molecular biology, and epidemiology of bacterial resistance. Microbiol Mol Biol Rev 65(2):232–260

48. Livermore DM (1995) Beta-lactamases in laboratory and clinical resistance. Clin Microbiol Rev 8(4):557–584

49. Jacobs L, Chenia HY (2007) Characterization of integrons and tetracycline resistance determinants in *Aeromonas spp.* isolated from South African aquaculture systems. Int J Food Microbiol 114(3):295–306

50. Smalla K, Van Overbeek L, Pukall R, Van Elsas J (1993) Prevalence of nptII and Tn5 in kanamycin-resistant bacteria from different environments. FEMS Microbiol Ecol 13(1): 47–58

51. Srinivasan V, Nam H, Nguyen L, Tamilselvam B, Murinda S, Oliver S (2005) Prevalence of antimicrobial resistance genes in *Listeria monocytogenes* isolated from dairy farms. Foodborne Pathog Dis 2(3):201–211

52. Zhu B (2007) Abundance dynamics and sequence variation of neomycin phosphotransferase gene (nptII) homologs in river water. Aquat Microb Ecol 48(2):131–140

53. Mao D, Yu S, Rysz M, Luo Y, Yang F, Li F, Hou J, Mu Q, Alvarez P (2015) Prevalence and proliferation of antibiotic resistance genes in two municipal wastewater treatment plants. Water Res 85:458–466

54. Colman BP, Arnaout CL, Anciaux S, Gunsch CK, Hochella MF Jr, Kim B, Lowry GV, McGill BM, Reinsch BC, Richardson CJ (2013) Low concentrations of silver nanoparticles in biosolids cause adverse ecosystem responses under realistic field scenario. PLoS One 8(2):e57189

55. Davies J, Davies D (2010) Origins and evolution of antibiotic resistance. Microbiol Mol Biol Rev 74(3):417–433

56. Holzem R, Stapleton H, Gunsch C (2014) Determining the ecological impacts of organic contaminants in biosolids using a high-throughput colorimetric denitrification assay: a case study with antimicrobial agents. Environ Sci Technol 48(3):1646–1655

57. Giger W, Alder AC, Golet EM, Kohler H-PE, McArdell CS, Molnar E, Siegrist H, Suter MJ-F (2003) Occurrence and fate of antibiotics as trace contaminants in wastewaters, sewage sludges, and surface waters. CHIMIA Int J Chem 57(9):485–491

58. Schwartz T, Kohnen W, Jansen B, Obst U (2003) Detection of antibiotic-resistant bacteria and their resistance genes in wastewater, surface water, and drinking water biofilms. FEMS Microbiol Ecol 43(3):325–335

59. Xi C, Zhang Y, Marrs CF, Ye W, Simon C, Foxman B, Nriagu J (2009) Prevalence of antibiotic resistance in drinking water treatment and distribution systems. Appl Environ Microbiol 75(17):5714–5718

60. Zhang X-X, Zhang T, Fang HH (2009) Antibiotic resistance genes in water environment. Appl Microbiol Biotechnol 82(3):397–414

61. Gardner CM, Gunsch CK (2017) Adsorption capacity of multiple DNA sources to clay minerals and environmental soil matrices less than previously estimated. Chemosphere 175:45–51

62. Gardner CM, Gwin CA, Gunsch CK (2018) A survey of crop derived transgenes in activated and digester sludges in wastewater treatment plants in the United States. Water Sci Technol 77(7–8):1810–1818

63. Ma B, Blackshaw RE, Roy J, He T (2011) Investigation on gene transfer from genetically modified corn (*Zea mays* L.) plants to soil bacteria. J Environ Sci Health B 46(7):590–599
64. Liu L, Li Y, Li S, Hu N, He Y, Pong R, Lin D, Lu L, Law M (2012) Comparison of next-generation sequencing systems. Biomed Res Int 2012:251364
65. Görg A, Weiss W, Dunn MJ (2004) Current two-dimensional electrophoresis technology for proteomics. Proteomics 4(12):3665–3685
66. Aebersold R, Mann M (2003) Mass spectrometry-based proteomics. Nature 422(6928):198
67. Caporaso JG, Kuczynski J, Stombaugh J, Bittinger K, Bushman FD, Costello EK, Fierer N, Pena AG, Goodrich JK, Gordon JI (2010) QIIME allows analysis of high-throughput community sequencing data. Nat Methods 7(5):335
68. Schloss PD, Larget BR, Handelsman J (2004) Integration of microbial ecology and statistics: a test to compare gene libraries. Appl Environ Microbiol 70(9):5485–5492
69. Glass EM, Wilkening J, Wilke A, Antonopoulos D, Meyer F (2010) Using the metagenomics RAST server (MG-RAST) for analyzing shotgun metagenomes. Cold Spring Harb Protoc 2010(1):pdb.prot5368
70. Wang S, Gunsch CK (2011) Effects of selected pharmaceutically active compounds on the ammonia oxidizing bacterium *Nitrosomonas europaea*. Chemosphere 82(4):565–572
71. Arnaout CL, Gunsch CK (2012) Impacts of silver nanoparticle coating on the nitrification potential of *Nitrosomonas europaea*. Environ Sci Technol 46(10):5387–5395
72. Andrieu C, De Freitas N, Doucet A, Jordan MI (2003) An introduction to MCMC for machine learning. Mach Learn 50(1–2):5–43
73. Goldberg DE, Holland JH (1988) Genetic algorithms and machine learning. Mach Learn 3(2):95–99
74. Guzdial M, Kolodner J, Hmelo C, Narayanan H, Carlson D, Rappin N, Hubscher R, Turns J, Newstetter W (1996) Computer support for learning through complex problem solving. Commun ACM 39(4):43–46
75. Kell DB (2006) Metabolomics, modelling and machine learning in systems biology–towards an understanding of the languages of cells. FEBS J 273(5):873–894
76. Alivisatos AP, Blaser M, Brodie EL, Chun M, Dangl JL, Donohue TJ, Dorrestein PC, Gilbert JA, Green JL, Jansson JK (2015) A unified initiative to harness Earth's microbiomes. Science 350(6260):507–508
77. Ikuma K, Holzem RM, Gunsch CK (2012) Impacts of organic carbon availability and recipient bacteria characteristics on the potential for TOL plasmid genetic bioaugmentation in soil slurries. Chemosphere 89(2):158–163
78. Gilbert JA, Jansson JK, Knight R (2014) The Earth Microbiome project: successes and aspirations. BMC Biol 12(1):69
79. Sheth RU, Cabral V, Chen SP, Wang HH (2016) Manipulating bacterial communities by in situ microbiome engineering. Trends Genet 32(4):189–200
80. Worley-Morse TO, Zhang L, Gunsch CK (2014) The long-term effects of phage concentration on the inhibition of planktonic bacterial cultures. Environ Sci: Processes Impacts 16(1):81–87
81. Worley-Morse TO, Deshusses MA, Gunsch CK (2015) Reduction of invasive bacteria in ethanol fermentations using bacteriophages. Biotechnol Bioeng 112(8):1544–1553
82. Morse TO, Morey SJ, Gunsch CK (2010) Microbial inactivation of *Pseudomonas putida* and *Pichia pastoris* using gene silencing. Environ Sci Technol 44(9):3293–3297

Courtney M. Gardner is an Assistant Professor of Civil and Environmental Engineering at Washington State University. Before she arrived at WSU, she was Postdoctoral Associate for the Nicholas School of the Environment and Pratt School of Engineering at Duke University. Her research investigates the interactions and dynamics between human mediated stressors on microbiomes in both natural and engineered systems, with a particular emphasis on the impacts of climate change on biological processes. Dr. Gardner is also interested in characterizing and applying the mechanisms driving microbial resilience to improve the efficiency of water treatment, bioremediation, and agricultural sustainability. She obtained her Ph.D. in Civil & Environmental Engineering from Duke for her work evaluating the potential contribution of transgenic crop biomass to environmental antibiotic resistance in wastewater treatment and agricultural systems. She also obtained her M.S. in Civil & Environmental Engineering at Duke and her B.S. in Biology from Stetson University. She has been recognized for her research and service with several awards, including a Graduate Research Fellowship from the National Science Foundation and the Jeffrey B. Taub Award from Duke University. She is a member of the ASCE Environmental Health and Water Quality committee and serves as a reviewer for several research journals, including *Bioresource Technology, Environmental Science: Water and Technology, Biodegradation,* and *Journal of Environmental Engineering.*

Claudia K. Gunsch is the Theodore S. Kennedy Associate Professor of Civil and Environmental Engineering and holds a secondary appointment in the Nicholas School of the Environment. She joined the Duke Faculty in 2004 after obtaining her PhD from the University of Texas at Austin, her MS from Clemson University, and her BS from Purdue University. Currently, she serves as the Director for IBIEM (*Integrative Bioinformatics for Investigating and Engineering Microbiomes*), a joint graduate training program between Duke and North Carolina A&T State University. She was selected as a Bass Fellow in 2016 and currently serves as a Faculty Mentor for the University Scholars Program as well as a Faculty Participant in the University Program in Ecology and the Center for Biomolecular and Tissue Engineering.

Her research bridges environmental engineering and molecular biotechnology. Current research foci include investigating the impacts of emerging contaminants on biological treatment processes, developing technologies for improving bioremediation efficacy, studying microbial evolution following exposure to anthropogenic contaminants, and developing innovative water treatment technologies. Her work has been funded by the National Science Foundation, US Environmental Protection Agency, National Institute for Environmental Health and Safety, as well as state-funding agencies and private industry. Dr. Gunsch was named a Fellow of the National Academy of Engineering for the US Frontiers of Engineering in 2011 and the Indo-American Frontiers of Engineering in 2014. In addition, she has been recognized for her research, teaching, and service activities with several awards including the 2009 National Science Foundation Faculty Early Career Development Award, 2011 Excellence in Review Award from *Environmental Science and Technology*, 2013 Langford Lectureship Award, 2016 Capers and Marion McDonald Award for Excellence in Mentoring and Advising, and the 2016 American Society of Civil Engineers (ASCE) Walter L. Huber Civil Engineering Research Prize.

She currently serves as an Associate Editor for *Biodegradation* and *Journal of Environmental Engineering*. She is also a member of the Editorial Board for *Industrial Biotechnology*. She has held several leadership roles within the Environmental & Water Resources Institute (EWRI) of ASCE as well as the Association of Environmental Engineering and Science Professors. Most recently, she was elected as the Vice Chair for the EWRI Environmental Council as well as the Chair of EWRI's Environmental Health and Water Quality Committee.

Chapter 8
Biofilms

Wen Zhang

Abstract This chapter focuses on biofilms, a common mode of bacteria growth ubiquitous in aquatic environments. Recent developments on biofilm detachment theories are summarized, and previously overlooked biofilm phenomena are explored, including the promising algal mats in natural environments, problematic distribution system biofilms in engineering environments, and the interaction between biofilms and emerging contaminants such as nanoparticles and disinfection byproducts. The goal of this chapter is to further demonstrate the importance of biofilms in environmental engineering and provide new insights for future biofilm control and utilization.

8.1 Introduction

Biofilms are entities of microorganisms enclosed within extracellular polymeric substances (EPS) and function as cooperative consortia. This mode of life is common to most microorganisms not only in the medical field but also in natural and engineered systems. Bacterial biofilms exist in multiple environments and can impact various aspects of people's daily lives. They can be beneficial to the field of water/wastewater treatment (biological nutrient removal) [1], but can be detrimental to multiple industries, such as food processing (biofouling) [2], health care (infection) [3], drinking water and ship hull maintenance (metal corrosion and pipe obstruction) [4, 5]. Most of these biofilms are comprised of complex and variable bacterial communities, which require effective biofilm control measures.

Biofilm control has two aspects: biofilm promotion and removal/prevention. Biofilm promotion is especially applicable in biological wastewater treatment, where sustained biofilm growth is desired [6]. Biofilm removal and prevention have been the focus of biofilm research in water distribution systems, dentistry, food processing procedures, and medical-related fields [7–10]. Lack of effective biofilm

W. Zhang (✉)
4190 Bell Engineering, University of Arkansas, Fayetteville, AR, USA
e-mail: wenzhang@uark.edu

© Springer Nature Switzerland AG 2020   135
D. J. O'Bannon (ed.), *Women in Water Quality*, Women in Engineering
and Science, https://doi.org/10.1007/978-3-030-17819-2_8

removal strategies can directly cause detrimental consequences, including biofouling, metal corrosion, pipe obstruction, and infection. The advances made in biofilm physiology and community complexity such as biofilm development theory [11] and quorum sensing [12] have greatly contributed to the understanding of biofilms and improved biofilm control strategies [13], but industries are still spending billions of dollars seeking effective biofouling control methods [14].

Scientists and engineers also struggle to explain many biofilm behaviors. For example, nutrient starvation was hypothesized and then confirmed as a biofilm detachment mechanism in *Pseudomonas aeruginosa*; however, the exact trigger for detachment was unclear [15]. On the another hand, even though quorum sensing is widely accepted as bacteria's way to regulate cell population density by gene expression [16], there is a need to identify its role in biofilm behavior in pure and mixed culture bacterial systems, as well as in different environmental conditions, since only a few investigations have been performed under mixed culture conditions [12]. As a result, fundamental studies of biofilm behavior are still lacking.

This chapter reviews recent development in biofilm studies, including mechanistic research on biofilm detachment, and biofilm applications in various engineering settings, with highlights in water and wastewater treatment. Lastly, interactions between biofilms and emerging contaminants will be explored. For the water and wastewater field, advances in biofilm understanding could translate to efficient contaminant removal in biofilm reactors, less biofouling on membrane surfaces, and safer drinking water delivered by transmission pipes. For other industries, these advances could mean reduced metal corrosion and pipe obstruction, less food waste due to package contamination, and more effective treatment for infectious diseases.

8.2   Biofilm Detachment

The five-stage development theory has been widely accepted for bacterial biofilm [17]. It includes:

- Stage 1—initial attachment: free-floating microorganisms attach to a surface with initial adhesion achieved through weak, reversible van der Waals forces.
- Stage 2—irreversible attachment: if reversible attached microorganisms do not separate from the surface, EPS begins to form to initiate irreversible attachment with a stronger adhesion force.
- Stage 3—maturation I: colonies develop an extracellular matrix comprised of various compounds to stabilize the biofilm and provide adhesion sites for other colonies.
- Stage 4—maturation II: biofilm grows as a structured combination of cells and EPS.
- Stage 5—dispersion: bacterial cells are released from the biofilm, some of which are able to reattach to the surface.

However, biofilm detachment is far from being well understood, unlike biofilm attachment. The interest in detachment processes primarily stemmed from removing or preventing biofilm in the industries mentioned above. Basic strategies include either killing biofilm microorganisms through biocides and disinfectants or chemical treatments that denature the EPS structure, depending on the specific application [18]. Although studies on these treatments shed light on biofilm detachment, they fail to offer a holistic view on the detachment process, which is also a part of the biofilm cycle—stage 5 dispersion.

Bryers [19] identified five categories of biofilm detachment: erosion, sloughing, human intervention, predator grazing, and abrasion. Erosion refers to single-cell or small cell cluster removal from the biofilm surface due to fluid shear stress, while sloughing is defined as the removal of a relatively large piece of biofilm from the surface. A later study by Telgmann et al. suggested sloughing is an integral part of biofilm development rather than a disturbance [20]. Human intervention, predator grazing, and abrasion result from other external forces can also exert on the biofilm surface. Mathematical models have been developed to characterize the detachment process [21–23], in which factors such as biofilm mass, thickness, species, growth rate, shear stress, substrate utilization, and oxygen depletion were considered.

A new theory was proposed to unify the existing studies on biofilm detachment: the differentiation between passive and active detachment. Passive biofilm detachment describes removal resulting from external forces, such as fluid shear stress and abrasion. This type of detachment occurs without a change in the biofilm metabolism [24, 25]. Active biofilm detachment describes removal resulting from internally driven mechanisms, including activation of a bacterial stress response mechanism [26], endogenous enzymatic degradation [27], and release of surface-binding proteins [28]. Environmental stressors, such as substrate limitations [29], oxygen depletion [30], and various chemical toxins [31], can cause internal changes including changes in biofilm structure (e.g., EPS destruction) and cell activity (e.g., respiration). This type of detachment depends on the species within the biofilm, given that each species has an independent stress response mechanism.

Fluid shear stress is one of the most recognized factors in passive detachment contributing to erosion and sloughing events [23, 24, 29, 32, 33]. Surface shear stress (τ) is defined by the following equation:

$$\tau(y) = \mu \, \partial u / \partial y$$

where μ is the dynamic viscosity of the fluid, u is the velocity of the fluid along the boundary, and y is the perpendicular distance to the support matrix. Flow rates can directly impact shear stress on the biofilm surface in a given flow chamber. In Huang et al. [25], two monoculture biofilms were developed in hollow fiber membrane-aerated bioreactors and subjected to various shear stresses measured by microparticle image velocimetry (μ-PIV) (Fig. 8.1). Positive correlations were observed between fluid-induced shear stress and detached biomass for both species. Other factors such as the biofilm growth stage (mature vs. immature) also plays an important role in detachment. In a similar experiment within an annular reactor, Choi and Margenroth [24] observed a similar trend (increasing detachment rate) during the

Flow Cell Wall

Fluid Shear

Biofilm

Membrane

Fig. 8.1 Schematic of passive detachment due to *fluid shear stress*. Biofilm develops on the exterior of a silicone hollow fiber membrane , and both live (green color) and dead (red color) bacterial cells are within the biofilm surrounded by EPS materials. When fluid shear stress is introduced to the biofilm surface, passive detachment happens without change of bacteria viability within *biofilms*. Detached particles include polymers (EPS and cell fragments), single cells (live and dead), and cell aggregates (a combination of cells and EPS). Higher fluid shear stress induces more biofilm detachment

initial increase in shear stress; however, the biofilm was able to adapt to the environment and restabilize the detachment. Based on the particle size analysis, they also concluded erosion was the dominant detachment mechanism under steady-state conditions, whereas sloughing was dominant following the sudden increase in shear stress.

One of the most common active detachment mechanisms is cell death, which can directly and indirectly cause biofilm detachment. Cell fragments from cell lysis contribute to biofilm detachment as a direct consequence; when the "glue"/EPS between bacteria and the surface ceases production due to cell death, biofilm detachment also follows. Various chemical toxins can change cell viability within a biofilm and lead to active detachment. Chelators such as ethylenediaminetetraacetic acid (EDTA) is one of the most commonly used chemical toxins to kill microorganisms and cause biofilm detachment [34, 35], as are various antibiotics, depending on the bacteria species [36]. Tobramycin is one of the most popular antibiotics for *Pseudomonas aeruginosa* infections. This compound is effective by binding bacterial ribosomes and preventing mRNA translation [36]. Although tobramycin is effective in killing suspended *P. aeruginosa* cells, a relatively high dosage (>1 mg/L) is required to inactivate cells within biofilms due to the resistance from EPS. Biofilm dispersion as well as cell death has been observed in *P. aeruginosa* biofilms (Fig. 8.2).

More biofilm detachment mechanisms are elucidated as quorum sensing theory develops, such as the effect of biosurfactant rhamnolipid production on *P. aeruginosa* biofilm and cyclic bis(3',5')guanylic acid (cyclic di-GMP) on *Shewanella oneidensis* MR-1 biofilm [37, 38]. Not only will these studies help determine the most effective biofilm removal strategy, they can also assist with the reactor design where beneficial biofilm is desired.

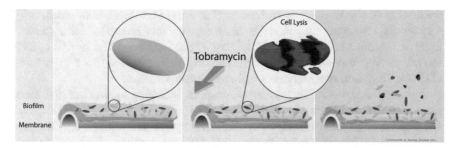

Fig. 8.2 Schematic of active detachment due to cell death. *P. aeruginosa* biofilm develops on the exterior of a silicone hollow fiber membrane, and both live (green color) and dead (red color) bacterial cells are within the biofilm surrounded by EPS materials. When the biofilm is exposed to antibiotic tobramycin, cell death is induced and cell lysis happens. The consequences include both active detachment and a decrease in cell viability within the biofilm structure. Detached particles include polymers (EPS and cell fragments), single cells (live and dead), and cell aggregates (a combination of cells and EPS). Particles with live cells tend to reattach to the surface again

8.3 Biofilms in Natural Environments

Biofilms are commonly found on sediments in natural aquatic environment, with the presence of a diverse group of microorganisms within the biofilm population [39]. Algal mats are a unique type of environmental biofilm that are often over-looked by environmental engineers.

Algal mats are microbial mats that exist in nature, in which layers of algae and cyanobacteria form biofilms on sediment surfaces, or as floating masses in water. Geologists investigated them in the 1970s, as calcified algal mats and trapped grains can form stromatolites, which have both geological and ecological significance [40]. A modern living mat can range from a thickness of millimeters to large algal mats which can develop in shallow-water estuaries with a thickness up to 10 cm [41]. These algal mats often include photosynthetic microorganisms, such as fila-mentous cyanobacteria, algae, and diatoms on the surface, and anaerobic sulfate-reducing bacteria (SRB) and methanogenic bacteria near the bottom [42, 43]. These communities can sequester organic materials and metals from the environment, while accumulating nutrients to sustain their own growth.

EPS plays an important role in algal mats. Cyanobacteria are generally recog-nized as the main EPS producers. Similar to bacterial biofilms, the composition of EPS produced by bacteria includes polysaccharides, proteins, lipids, nucleic acids, and humic substances [44]. The EPS matrix fulfills many functions within algal mats [45], including binding of dissimilar communities and formation of a three-dimensional structure to allow air/water access; physical stabilization of microbial cells under variable hydrodynamic regimes; and helping the microbial mat com-munity resist multiple stress conditions, such as nutrient shortages, UV exposure, or desiccation. EPS can either promote or inhibit carbon precipitation depending on the amount of negatively charged functional groups in the microenvironment [43]. Specifically, the negatively charged acidic groups within EPS can bind free calcium ions from solution to inhibit the carbonate precipitation, or the degradation of EPS can release free calcium ions to promote carbonate precipitation. However, the role of EPS in the initial attachment of algal mats remains unclear.

Algal mats can be used for various applications, including aquaculture (fish feeding and effluent treatment), bioremediation (heavy metal and radionuclide removal and organic degradation), agriculture (soil aggregation and nutrient enrichment), and bioenergy (biohydrogen production) [46, 47]. Bioremediation is the most popular application of algal mats. Kalin et al. used a three-step process to remove uranium from mining wastewater including sequestration of uranium (U) by algae, removal of U-algal particulates by precipitation, and reducing U(VI) to U(IV) in precipitates [48]. Bender et al. used stratified algal mats to transform nitrogenous wastes into cellular protein in marine-cultured wastewater. These mats rapidly metabolized other fish wastes while providing excess oxygen for the nitrifying bacteria to remove ammonia [49]. Das et al. found various species of algae and fungi which lived symbiotically in an acid mine drainage environment, where algae produced an anoxic zone for SRB and helped in biogenic alkalinity generation, while fungi absorbed a significant amount of metals in their cell wall or EPS [50]. Al-Thukair et al. investigated a microbial community of algal mats in an oil-polluted coast of Saudi Arabia and found that the cyanobacteria were all well adapted to the oil pollution and developed a tolerance to the high temperature, salinity, and desiccation periods [51]. Nzengung et al. observed sequestration and transformation of chlorinated and brominated organic contaminants by algal mats in a sealed microbial mat bioreactor [52]. Paniagua-Michel and Garcia used constructed microbial mats to effectively remove ammonia and nitrate from shrimp culture effluent [53]. Safonova et al. used selected algal-bacterial consortia to treat industrial wastewater and observed a significant decrease in contaminants such as biochemical oxygen demand (BOD), phenols, anionic surface-active substances, oil spills, copper, nickel, zinc, manganese, and iron [54]. Jacques and McMartin evaluated the role of algae and algal mats in reducing light extractable petroleum hydrocarbon concentrations in petroleum-contaminated water [55]. Surprisingly, however, very few studies focused on domestic wastewater treatment using algal mats.

Algal-bacterial processes have shown success in treating hazardous contaminants [56], but their interaction with other emerging contaminants such as endocrine-disrupting compounds and nanoparticles in wastewater is still unknown. Algae have shown promise in removing these contaminants: bioaccumulation and degradation of endocrine disrupters were studied with algae [57, 58], and interactions between silver and gold nanoparticles and freshwater algae were investigated in a few studies [59, 60]. Mixed culture biofilms have also shown promise in nanoparticle removal, due to their enhanced resistance to nanoparticle toxicity [61] as well as their affinity for sorption [62]. These studies indicate algal-bacterial biofilms (such as algal mats) have a favorable chance of success for nanoparticle removal while maintaining their resistance to the nanoparticle toxicity; however, more studies are needed. Overall, algal mats could be an effective treatment tool for contaminant removal, compared to heterotrophic bacteria biofilms.

8.4 Biofilms in Engineering Environments

Biofilms have been widely applied in the field of wastewater treatment. Biofilm reactors such as trickling filters and rotating biological contactors have been commonly used in the industrial waste and municipal sectors due to their compact design, low quantities of sludge production, and resistance to shock loads [63]. New technologies such as membrane-aerated biofilm reactors and fluidized-bed bioreactors have re-emerged as viable options for municipal wastewater treatment [64, 65]. Biofilms, the key player within these treatment methods, is effective at removing traditional contaminants such as nutrients as well as trace contaminants.

Researchers have started investigating biofilms in drinking water treatment process in recent decades. Biofilms have been found in treatment units such as sedimentation basins [66], biological filters [67], and drinking water distribution systems (DWDSs) [68]. Biofilms in DWDSs has gained special interest, as it could have a direct impact on consumer health. In the United States, secondary disinfectants such as free chlorine and chloramines are applied in DWDSs to help prevent bacterial regrowth and ensure drinking water is adequately disinfected at the tap [69]. However, despite the use of secondary disinfectants, biofilms in DWDSs are ubiquitous. Biofilms are more resistant to disinfectants than suspended bacteria and have been found to harbor 25 times more bacterial cells than the adjacent bulk water in the DWDS [70]. Numerous studies have investigated the various biofilm-related issues in water treatment systems, including the microbial diversity of the filter biofilm [71], biofilm resistance to disinfectants [72], and biofilm growth on pipe materials [73]. However, DWDS biofilms vary with water quality, water age, disinfectants used, pipe material, and other environmental factors. Most studies used simulated DWDSs, and the biofilms formed might not represent the real-world sce-

Fig. 8.3 Left photo shows the cross section of a 4-inch cast iron pipe taken off service from the Fayetteville DWDS (AR); right photo shows the close-up shot of biofilms formed within the same pipe

narios. Fig. 8.3 shows a section of the DWDS pipe taken off service from the city of Fayetteville (AR) in September, 2015. Visible formation of biofilms (tubercles) covered the interior surface of the 4-inch cast iron pipe. The composition and bacterial species within the biofilm are still unknown, which poses an immediate health threat to consumers.

Interactions between biofilm and trace elements within DWDSs are also of interest. The growth of distribution biofilms relies on the presence of electron donors, such as biodegradable organic carbon and nitrogen species. The adsorption of certain contaminants in the treated water depends heavily on the biological activity inside the pipes, such as the production of nitrate from nitrifying organisms [74]. These biofilms may also have a sorptive capacity and thus are potentially significant sinks for contaminants such as metals [75]. Templeton's study investigated the distribution of aqueous Pb(II) sorbed at the interface between *Burkholderia cepacia* biofilms and hematite or corundum surfaces and found the formation of a monolayer biofilm did not affect reactive sites on the metal oxide surfaces, and significant Pb biofilm uptake was observed [75]. Ancion et al. investigated the absorption rates of zinc, copper, and lead in freshwater biofilms and observed rapid metal accumulation by the biofilm. When the exposed biofilm was transferred to uncontaminated water, they retained the accumulated metals for at least 14 days [76]. The changes in the microbial community structure during exposure to metals as well as the EPS composition of biofilms have also been studied [77].

The use of biofilms to capture trace elements could have profound implications on modern-day water catastrophes, such as the recent lead crisis in Flint, Michigan. The Flint drinking water became contaminated with lead in April 2014. The problem originated from the cost-cutting move to use the Flint River as its drinking water source. From the plant operation perspective, the lack of appropriate corrosion control resulted in the leaching of lead from pipes and fixtures into the drinking water. Although most recently exposed in Flint, lead contamination in drinking water can be a US-wide problem, since nearly all homes built prior to the 1980s have lead solder connecting copper pipes and many US cities still have a majority of lead pipes in their distribution systems transmitting drinking water. Trace amounts of lead can cause serious health issues and can severely affect mental and physical development of children under the age of 6. The Maximum Contamination Level Goal (MCLG) for lead is zero in the EPA primary drinking water standards.

The correction of lead contamination in drinking water is no easy task. Researchers have been investigating the impact of partial line replacement. Unfortunately, it was shown that partial replacement of lead service lines could exacerbate the problem by releasing higher levels of lead into the delivered water [78]. To make matters worse, lead-containing parts were also used in premise plumbing, where replacement cost will likely fall directly onto the consumers themselves. It is critical to identify other possible sources of such contamination, and their interactions with biofilms in DWDSs.

8.5 Biofilms and Emerging Contaminants

Water reuse practices have increased to alleviate the stress from diminishing freshwater resources [79], but they bring concerns over emerging contaminants, including nanomaterials, perfluorinated compounds, pharmaceuticals, endocrine-disrupting compounds, drinking water disinfection byproducts (DBPs), pesticides, etc. [80]. In general, biofilms have been found to resist the toxicity of various contaminants better than their planktonic counterparts due to the protection effect of EPS [81], and to accumulate these contaminants at different concentrations [82, 83]. However, not all biofilm-contaminant interactions bring beneficial impacts.

N-nitrosamines are a highly toxic and non-halogenated group of disinfection byproducts (DBPs), primarily associated with chloramination [84]. The nationwide occurrence data for N-nitrosamines has identified seven separate species in EPA Method 521, with N-nitrosodimethylamine (NDMA) being the most prevalent [85]. NDMA was measured in six distribution systems out of 20 public water utilities in Canada above the detection limit of 5 ng/L [86]. Multiple studies showed increasing NDMA concentrations with increasing water age within distribution systems [87, 88].

Known NDMA precursors include quaternary amine-containing coagulants and anion-exchange resins and wastewater-impacted source waters containing pharmaceuticals and personal care products [84]. Biofilm EPS contains organic materials that are comprised of chemical moieties similar to known N-nitrosamine precursors (e.g., secondary amines [89]). Exopolysaccharides, a component of EPS that bonds cells together and facilitates microcolony formation, are produced and accumulated by attached cells during biofilm formation [90, 91]. Alginate, a major component of the exopolysaccharides in *Pseudomonas aeruginosa* biofilm, plays an important role during biofilm development [92]. The gene responsible for exopolysaccharide alginate synthesis is 3–5 times more active in attached growth systems than suspended cells [93, 94]. *P. aeruginosa* is pathogenic and is sometimes found to cause periodic contamination in DWDSs [95]. Although alginate has been widely associated with *P. aeruginosa* bacterial biofilm, it can also be biosynthesized in other bacteria strains such as *Pseudomonads* and *Azotobacter* spp., as well as algal and cyanobacteria species [96, 97]. Seo's research group used *Pseudomonas* strains to produce different quantities and composition of EPS and extracted bacterial EPS from pure and mixed culture biofilm and chlorinated the extracts [98, 99]. Results showed that proteins within the biofilm EPS contributed to the formation of haloacetonitriles, a group of nitrogen-containing DBPs, presumably through substitution reactions with amino acids containing unsaturated organic carbon or conjugated bonds in the R-group. It is likely that biofilm exopolysaccharides could contribute to the formation of nitrogen-containing DBPs.

Polysaccharide intercellular adhesion (PIA), another well-characterized biofilm exopolysaccharide, is also called poly-N-acetyl-glucosamine (or PNAG), which is a common component of EPS produced by *Escherichia coli* [100, 101], *Staphylococcus aureus*, and *S. epidermidis* [102]. For DBP formation, bacteria species relevant in DWDSs are extremely important, such as *E. coli* (indicator bacteria). The composition and unique structure of PIA/PNAG is informative in the formation of nitrogen-containing DBPs. PIA/PNAG contains functional groups that resemble secondary and tertiary amines known to react with chloramines to form NDMA [89]. An important feature of the PIA/PNAG molecule is its partial deacetylation, which creates free amino groups. This provides additional evidence that biofilm exopolysaccharides could contribute to the formation of N-nitrosamines. Studies are currently underway to verify this reaction mechanism.

In addition to acting as DBP precursors, biofilms in DWDSs can degrade water quality by consuming disinfectant residual, enabling survival of potential pathogens, promoting nitrification, and accelerating metal corrosion [103–106]. Traditional biofilm removal strategy such as a chlorine burn may no longer be suitable, considering the potential formation of toxic DBPs. As a result, it is even more urgent to redesign biofilm control measures, especially for water pipelines where no additional barriers exist before water reaches consumers.

In summary, this chapter presented both beneficial and detrimental impacts biofilms could have in natural and engineering environments. While the manipulation of biofilm composition and bacterial gene expression can be leveraged to achieve more effective contaminants removal, the accompanying consequences should also be considered. For example, if lead leached into drinking water can be accumulated within DWDS biofilms, it is also possible lead can be subsequently released from these biofilms. In fact, the wastewater treatment sector is already dealing with similar effect, where nanoparticles adsorbed into activated sludge negatively impact the subsequent processes such as anaerobic digestion [107]. Therefore, a total life cycle approach should be considered whether biofilm is accounted as a culprit or applied as a potential solution to an engineering problem.

References

1. Lazarova V, Manem J (1995) Biofilm characterization and activity analysis in water and wastewater treatment. Water Res 29(10):2227–2245
2. Blackman IC, Frank JF (1996) Growth of *Listeria monocytogenes* as a biofilm on various food-processing surfaces. J Food Prot® 59(8):827–831
3. Donlan RM (2001) Biofilm formation: a clinically relevant microbiological process. Clin Infect Dis 33(8):1387–1392
4. LeChevallier MW, Babcock TM, Lee RG (1987) Examination and characterization of distribution system biofilms. Appl Environ Microbiol 53(12):2714–2724
5. Schultz M et al (2011) Economic impact of biofouling on a naval surface ship. Biofouling 27(1):87–98
6. Gross M et al (2013) Development of a rotating algal biofilm growth system for attached microalgae growth with in situ biomass harvest. Bioresour Technol 150(0):195–201
7. Marsh PD, Bradshaw DJ (1995) Dental plaque as a biofilm. J Ind Microbiol Biotechnol 15(3):169–175
8. Piriou PH et al (1997) Prevention of bacterial growth in drinking water distribution systems. Water Sci Technol 35(11):283–288
9. Raad II, Darouiche RO (1996) Catheter-related septicemia: risk reduction. Inf Med 13:807–816
10. Zottola EA, Sasahara KC (1994) Microbial biofilms in the food processing industry--should they be a concern? Int J Food Microbiol 23(2):125–148

11. Stoodley P et al (2002) Biofilms as complex differentiated communities. Annu Rev Microbiol 56(1):187–209
12. Shrout JD, Nerenberg R (2012) Monitoring bacterial twitter: does quorum sensing determine the behavior of water and wastewater treatment biofilms? Environ Sci Technol 46(4):1995–2005
13. Xavier JB et al (2005) Biofilm-control strategies based on enzymic disruption of the extracellular polymeric substance matrix–a modelling study. Microbiology 151(Pt 12):3817–3832
14. Callow M E, Callow J A (2002). Marine biofouling: a sticky problem. Biologist, 49(1):1–5.
15. Hunt SM et al (2004) Hypothesis for the role of nutrient starvation in biofilm detachment. Appl Environ Microbiol 70(12):7418–7425
16. Miller MB, Bassler BL (2001) Quorum sensing in bacteria. Annu Rev Microbiol 55(1):165–199
17. Sauer K et al (2002) *Pseudomonas aeruginosa* displays multiple phenotypes during development as a biofilm. J Bacteriol 184(4):1140–1154
18. Chen X, Stewart PS (2000) Biofilm removal caused by chemical treatments. Water Res 34(17):4229–4233
19. Bryers JD (1988) Modeling biofilm accumulation. Physiol Model Microbiol 2:109–144
20. Telgmann U, Horn H, Morgenroth E (2004) Influence of growth history on sloughing and erosion from biofilms. Water Res 38(17):3671–3684
21. Chaudhry MAS, Beg SA (1998) A review on the mathematical modeling of biofilm processes: advances in fundamentals of biofilm modeling. Chem Eng Technol 21(9):701–710
22. Luna E et al (1996) Detachment and diffusive-convective transport in an evolving heterogeneous two-dimensional biofilm hybrid model. Phys Rev E 70(061909):1–8
23. Picioreanu C, van Loosdrecht MCM, Heijnen JJ (2001) Two-dimensional model of biofilm detachment caused by internal stress from liquid flow. Biotechnol Bioeng 72(2):205–218
24. Choi YC, Morgenroth E (2003) Monitoring biofilm detachment under dynamic changes in shear stress using laser-based particle size analysis and mass fractionation. Water Sci Technol J Int Assoc Water Pollut Res 47(5):69–76
25. Huang Z et al (2013) Shear-induced detachment of biofilms from hollow fiber silicone membranes. Biotechnol Bioeng 110(2):525–534
26. Zhang W et al (2014) Glutathione-gated potassium efflux as a mechanism of active biofilm detachment. Water Environ Res 86(5):462–469
27. Lee SF, Li YH, Bowden GH (1996) Detachment of Streptococcus mutans biofilm cells by an endogenous enzymatic activity. Infect Immun 64(3):1035–1038
28. Bastian FO, Elzer PH, Wu XC (2012) Spiroplasma spp. biofilm formation is instrumental for their role in the pathogenesis of plant, insect and animal diseases. Exp Mol Pathol 93(1):116–128
29. Peyton BM, Characklis WG (1993) A statistical analysis of the effect of substrate utilization and shear stress on the kinetics of biofilm detachment. Biotechnol Bioeng 41(7):728–735
30. Thormann KM et al (2005) Induction of rapid detachment in Shewanella oneidensis MR-1 biofilms. J Bacteriol 187(3):1014–1021
31. Hsieh KM et al (1994) Interactions of microbial biofilms with toxic trace metals: 1. Observation and modeling of cell growth, attachment, and production of extracellular polymer. Biotechnol Bioeng 44(2):219–231
32. Stewart PS (1993) A model of biofilm detachment. Biotechnol Bioeng 41:111–117
33. Bakke R et al (1990) Modeling a monopopulation biofilm system: *Pseudomonas aeruginosa*. In: Biofilms. Wiley, New York, pp 487–520
34. Webb JS et al (2003) Cell death in *Pseudomonas aeruginosa* biofilm development. J Bacteriol 185(15):4585–4592
35. Banin E, Brady KM, Greenberg EP (2006) Chelator-induced dispersal and killing of *Pseudomonas aeruginosa* cells in a biofilm. Appl Environ Microbiol 72(3):2064–2069
36. Walters MC III et al (2003) Contributions of antibiotic penetration, oxygen limitation, and low metabolic activity to tolerance of *Pseudomonas aeruginosa* biofilms to ciprofloxacin and tobramycin. Antimicrob Agents Chemother 47(1):317–323

37. Boles BR, Thoendel M, Singh PK (2005) Rhamnolipids mediate detachment of *Pseudomonas aeruginosa* from biofilms. Mol Microbiol 57(5):1210–1223
38. Thormann KM et al (2006) Control of formation and cellular detachment from Shewanella oneidensis MR-1 biofilms by cyclic di-GMP. J Bacteriol 188(7):2681–2691
39. Meyer-Reil L-A (1994) Microbial life in sedimentary biofilms—the challenge to microbial ecologists. Mar Ecol Prog Ser 112:303–311
40. Riding R (2000) Microbial carbonates: the geological record of calcified bacterial–algal mats and biofilms. Sedimentology 47(s1):179–214
41. Escartın J, Aubrey DG (1995) Flow structure and dispersion within algal mats. Estuar Coast Shelf Sci 40(4):451–472
42. Ward DM (1978) Thermophilic methanogenesis in a hot-spring algal-bacterial mat (71 to 30 degrees C). Appl Environ Microbiol 35(6):1019–1026
43. Dupraz C et al (2009) Processes of carbonate precipitation in modern microbial mats. Earth Sci Rev 96(3):141–162
44. Nielsen PH, Jahn A, Palmgren R (1997) Conceptual model for production and composition of exopolymers in biofilms. Water Sci Technol 36(1):11–19
45. Decho AW (1990) Microbial exopolymer secretions in ocean environments: their role (s) in food webs and marine processes. Oceanogr Mar Biol Annu Rev 28(7):73–153
46. Bender J, Phillips P (2004) Microbial mats for multiple applications in aquaculture and bio-remediation. Bioresour Technol 94(3):229–238
47. Roeselers G, Van Loosdrecht MCM, Muyzer G (2008) Phototrophic biofilms and their potential applications. J Appl Phycol 20(3):227–235
48. Kalin M, Wheeler WN, Meinrath G (2004) The removal of uranium from mining waste water using algal/microbial biomass. J Environ Radioact 78(2):151–177
49. Bender J et al (2004) A waste effluent treatment system based on microbial mats for black sea bass *Centropristis striata* recycled-water mariculture. Aquac Eng 31(1):73–82
50. Das BK et al (2009) Occurrence and role of algae and fungi in acid mine drainage environment with special reference to metals and sulfate immobilization. Water Res 43(4):883–894
51. Al-Thukair AA, Abed RMM, Mohamed L (2007) Microbial community of cyanobacteria mats in the intertidal zone of oil-polluted coast of Saudi Arabia. Mar Pollut Bull 54(2):173–179
52. Nzengung VA et al (2003) Sequestration and transformation of water soluble halogenated organic compounds using aquatic plants, algae, and microbial mats. In: Phytoremediation: transformation and control of contaminants. Wiley-Interscience, Hoboken, pp 497–528
53. Paniagua-Michel J, Garcia O (2003) Ex-situ bioremediation of shrimp culture effluent using constructed microbial mats. Aquac Eng 28(3):131–139
54. Safonova E et al (2004) Biotreatment of industrial wastewater by selected algal-bacterial consortia. Eng Life Sci 4(4):347–353
55. Jacques NR, McMartin DW (2009) Evaluation of algal phytoremediation of light extractable petroleum hydrocarbons in subarctic climates. Remediat J 20(1):119–132
56. Muñoz R, Guieysse B (2006) Algal–bacterial processes for the treatment of hazardous contaminants: a review. Water Res 40(15):2799–2815
57. Sethunathan N et al (2004) Algal degradation of a known endocrine disrupting insecticide, α-endosulfan, and its metabolite, endosulfan sulfate, in liquid medium and soil. J Agric Food Chem 52(10):3030–3035
58. Geyer HJ et al (2000) Bioaccumulation and occurrence of endocrine-disrupting chemicals (EDCs), persistent organic pollutants (POPs), and other organic compounds in fish and other organisms including humans. In: Bioaccumulation–new aspects and developments. Springer, Berlin/New York, pp 1–166
59. He D, Dorantes-Aranda JJ, Waite TD (2012) Silver nanoparticle-algae interactions: oxidative dissolution, reactive oxygen species generation and synergistic toxic effects. Environ Sci Technol 46(16):8731–8738
60. Renault S et al (2008) Impacts of gold nanoparticle exposure on two freshwater species: a phytoplanktonic alga (*Scenedesmus subspicatus*) and a benthic bivalve (*Corbicula fluminea*). Gold Bull 41(2):116–126

61. Sheng Z, Liu Y (2011) Effects of silver nanoparticles on wastewater biofilms. Water Res 45(18):6039–6050
62. Hou L et al (2012) Removal of silver nanoparticles in simulated wastewater treatment processes and its impact on COD and NH 4 reduction. Chemosphere 87(3):248–252
63. Metcalf L, Eddy HP, Tchobanoglous G (1972) Wastewater engineering: treatment, disposal, and reuse. McGraw-Hill, New York
64. Şen S, Demirer G (2003) Anaerobic treatment of real textile wastewater with a fluidized bed reactor. Water Res 37(8):1868–1878
65. Pankhania M, Stephenson T, Semmens MJ (1994) Hollow fibre bioreactor for wastewater treatment using bubbleless membrane aeration. Water Res 28(10):2233–2236
66. Walden C, Carbonero F, Zhang W (2015) Preliminary assessment of bacterial community change impacted by chlorine dioxide in a water treatment plant. J Environ Eng 142(2):04015077
67. Lautenschlager K et al (2014) Abundance and composition of indigenous bacterial communities in a multi-step biofiltration-based drinking water treatment plant. Water Res 62:40–52
68. Berry D, Xi C, Raskin L (2006) Microbial ecology of drinking water distribution systems. Curr Opin Biotechnol 17(3):297–302
69. Seidel CJ et al (2005) Have utilities switched to chloramines? J Am Water Works Assoc 97(10):87–97
70. Servais P et al (2004) Biofilm in the Parisian suburbs drinking water distribution system. J Water Supply Res Technol 53(5):313–324
71. Pinto AJ, Xi CW, Raskin L (2012) Bacterial community structure in the drinking water microbiome is governed by filtration processes. Environ Sci Technol 46(16):8851–8859
72. Schwartz T, Hoffmann S, Obst U (2003) Formation of natural biofilms during chlorine dioxide and UV disinfection in a public drinking water distribution system. J Appl Microbiol 95(3):591–601
73. Niquette P, Servais P, Savoir R (2000) Impacts of pipe materials on densities of fixed bacterial biomass in a drinking water distribution system. Water Res 34(6):1952–1956
74. van der Wielen PW, Voost S, van der Kooij D (2009) Ammonia-oxidizing bacteria and archaea in groundwater treatment and drinking water distribution systems. Appl Environ Microbiol 75(14):4687–4695
75. Templeton AS et al (2001) Pb(II) distributions at biofilm-metal oxide interfaces. Proc Natl Acad Sci U S A 98(21):11897–11902
76. Ancion P-Y, Lear G, Lewis GD (2010) Three common metal contaminants of urban runoff (Zn, Cu & Pb) accumulate in freshwater biofilm and modify embedded bacterial communities. Environ Pollut 158(8):2738–2745
77. Jang A et al (2001) Effect of heavy metals (Cu, Pb, and Ni) on the compositions of EPS in biofilms. Water Sci Technol 43(6):41–48
78. St. Clair J et al (2016) Long-term behavior of simulated partial lead service line replacements. Environ Eng Sci 33(1):53–64
79. Miller GW (2006) Integrated concepts in water reuse: managing global water needs. Desalination 187(1–3):65–75
80. Richardson SD (2009) Water analysis: emerging contaminants and current issues. Anal Chem 81(12):4645–4677
81. Walden C, Zhang W (2016) Biofilms versus activated sludge: considerations in metal and metal oxide nanoparticle removal from wastewater. Environ Sci Technol 50(16):8417–8431
82. Zhang W et al (2009) Accumulation of tetracycline resistance genes in aquatic biofilms due to periodic waste loadings from swine lagoons. Environ Sci Technol 43:7643–7650
83. Munoz, G et al (2018). Spatio-temporal dynamics of per and polyfluoroalkyl substances (PFASs) and transfer to periphytic biofilm in an urban river: case-study on the River Seine. Environmental Science and Pollution Research 25(24):23574–23582.
84. Krasner SW et al (2013) Formation, precursors, control, and occurrence of nitrosamines in drinking water: a review. Water Res 47(13):4433–4450

85. Dai N, Mitch WA (2013) Relative importance of N-nitrosodimethylamine compared to total N-nitrosamines in drinking waters. Environ Sci Technol 47(8):3648–3656
86. Charrois JWA et al (2007) Occurrence of N-nitrosamines in Alberta public drinking-water distribution systems. J Environ Eng Sci 6(1):103–114
87. Wilczak A et al (2003) Formation of NDMA in chloraminated water coagulated with DADMAC cationic polymer. J Am Water Works Assoc 95(9):94–106
88. Barrett S et al (2003) Occurrence of NDMA in drinking water: a North American survey, 2001–2002. In: American Water Works Association annual conference. AWWA, Anaheim
89. Lee C et al (2007) Oxidation of N-nitrosodimethylamine (NDMA) precursors with ozone and chlorine dioxide: kinetics and effect on NDMA formation potential. Environ Sci Technol 41(6):2056–2063
90. Gerke C et al (1998) Characterization of the N-acetylglucosaminyltransferase activity involved in the biosynthesis of the *Staphylococcus epidermidis* polysaccharide intercellular adhesin. J Biol Chem 273(29):18586–18593
91. Cramton SE et al (1999) The intercellular adhesion (ica) locus is present in Staphylococcus aureus and is required for biofilm formation. Infect Immun 67(10):5427–5433
92. Cotton L, Graham R, Lee R (2009) The role of alginate in *P. aeruginosa* PAO1 biofilm structural resistance to gentamicin and ciprofloxacin. J Exp Microbiol Immunol 13:58–62
93. Davies DG, Geesey GG (1995) Regulation of the alginate biosynthesis gene ALGC in *Pseudomonas aeruginosa* during biofilm development in continuous-culture. Appl Environ Microbiol 61(3):860–867
94. Davies DG, Chakrabarty AM, Geesey GG (1993) Exopolysaccharide production in biofilms – substratum activation of alginate gene-expression by *Pseudomonas aeruginosa*. Appl Environ Microbiol 59(4):1181–1186
95. Parsek MR, Singh PK (2003) Bacterial biofilms: an emerging link to disease pathogenesis. Annu Rev Microbiol 57:677–701
96. Gacesa P (1998) Bacterial alginate biosynthesis – recent progress and future prospects. Microbiology-UK 144:1133–1143
97. Sutherland IW (2001) Biofilm exopolysaccharides: a strong and sticky framework. Microbiology-UK 147:3–9
98. Wang ZK, Kim J, Seo Y (2012) Influence of bacterial extracellular polymeric substances on the formation of carbonaceous and nitrogenous disinfection Byproducts. Environ Sci Technol 46(20):11361–11369
99. Wang ZK, Choi O, Seo Y (2013) Relative contribution of biomolecules in bacterial extracellular polymeric substances to disinfection byproduct formation. Environ Sci Technol 47(17):9764–9773
100. Cerca N, Jefferson KK (2008) Effect of growth conditions on poly-N-acetylglucosamine expression and biofilm formation in *Escherichia coli*. FEMS Microbiol Lett 283(1):36–41
101. Izano EA et al (2007) Poly-N-acetylglucosamine mediates biofilm formation and antibiotic resistance in Actinobacillus pleuropneumoniae. Microb Pathog 43(1):1–9
102. Joo HS, Otto M (2012) Molecular basis of in vivo biofilm formation by bacterial pathogens. Chem Biol 19(12):1503–1513
103. Camper AK (2004) Involvement of humic substances in regrowth. International Journal of Food Microbiology 92(3):355–364
104. Emtiazi F et al (2004) Investigation of natural biofilms formed during the production of drinking water from surface water embankment filtration. Water Research, 38(5):1197–1206
105. Teng F, Guan Y, Zhu W (2008) Effect of biofilm on cast iron pipe corrosion in drinking water distribution system: Corrosion scales characterization and microbial community structure investigation. Corrosion Science 50(10):2816–2823
106. Zhang Y, Edwards M (2009) Accelerated chloramine decay and microbial growth by nitrification in premise plumbing. American Water Works Association.Journal 101(11):51–62
107. Yang Y, Zhang C, Hu Z (2013) Impact of metallic and metal oxide nanoparticles on wastewater treatment and anaerobic digestion. Environ Sci: Processes Impacts 15(1):39–48

Wen Zhang is an Associate Professor in the Department of Civil Engineering at the University of Arkansas (UA). She is a registered Professional Engineer in the State of Arkansas. She teaches undergraduate and graduate courses including Environmental Engineering, Hydraulics, Microbiology for Environmental Engineers, and Water Reuse. Her research goal is to improve wastewater and drinking water treatment technologies using microbial processes. She conducts fundamental and applied research centering on biofilm processes, such as biofilm detachment mechanisms, interactions between biofilm and emerging contaminants, and bacterial communities in water and wastewater treatment. In addition to teaching and research, she also takes an active role serving as the faculty advisor in the UA chapter of Society of Women Engineers (SWE) and in the Women in Academia committee within national SWE.

Dr. Zhang graduated in 2006 with a Bachelor's degree in Environmental Engineering from Tongji University in Shanghai, China. She experienced water scarcity firsthand due to limited resources and a rapidly growing population and immediately realized the importance of water research. Since then, from being a graduate student at the University of Kansas to being a faculty member at the UA, she has been positively influenced and guided by role models and mentors. They were instrumental to her personal and professional growth. Prof. Belinda Sturm at the University of Kansas advised Dr. Zhang's MS thesis in studying the fate of antibiotic-resistance genes in surface water and biofilms receiving agricultural runoff. Prof. Sturm opened the door of the fascinating research world and led her to all the opportunities and challenges in environmental engineering. Prof. M. Kathy Banks prepared Dr. Zhang for an academic career by both directing her doctoral dissertation in biofilm detachment mechanisms and sharing her leadership experiences at Purdue University. Prof. Heather Nachtmann at the UA has been instrumental to Dr. Zhang's tenure and promotion process since she arrived at the UA in 2011. Dr. Zhang is thankful for having all these strong female mentors who sustain successful careers while being busy mothers.

Dr. Zhang has been awarded funding from the National Science Foundation, the National Institutes for Water Resources, the Water Research Foundation, and local water utilities. This book chapter is intended to share the research progress made by Dr. Zhang with the readers, and more importantly, to inspire and attract young talent into the field of environmental research. New and advanced technologies that can provide clean and safe water are desired more than ever in the urgency of water reuse.

Part III
Water Quality in Natural Systems

Chapter 9
The Microbial Ecology and Bioremediation of Chlorinated Ethene-Contaminated Environments

Jennifer G. Becker

Abstract The microbial ecology of tetrachloroethene (PCE)- and trichloroethene (TCE)-contaminated sites is complex. Fundamentally, accurate prediction of contaminant fate, the survival of dehalorespiring populations, and, thus, the performance of engineered bioremediation approaches at these sites are feasible only if the correct kinetic models are applied, and meaningful and mathematically independent parameter estimates are input into these models. A model that incorporates biomass inactivation at high chlorinated ethene concentrations, as well as the self-inhibitory and competitive inhibition effects that the elevated chlorinated ethene concentrations exert on dechlorination reactions, must be utilized to accurately predict dehalorespiring population substrate interactions and growth. The initial conditions used in batch laboratory kinetic assays, including the initial limiting substrate (S_0)-to-initial biomass concentration (X_0) ratio and the S_0-to-half-saturation constant (K_S) ratio, must be carefully selected to ensure that the parameter estimates are meaningful and independent. Kinetic assays conducted at appropriate S_0/X_0 and S_0/K_S ratios suggest that the substrate utilization kinetics of many PCE-to-dichloroethene (DCE) dehalorespirers are faster than those of *Dehalococcoides mccartyi* strains. Integration of mathematical simulations using appropriate dehalorespiration models and dehalorespiring co-culture experiments also showed that PCE-to-DCE dehalorespirers tend to outcompete *D. mccartyi* strains for higher chlorinated ethenes. Where dense nonaqueous-phase liquid (DNAPL) contamination is present, the fast substrate utilization kinetics of PCE-to-DCE dehalorespirers allow them to grow close to the DNAPL-water interface and control dissolution bioenhancement. Under excess electron donor conditions, *D. mccartyi* strains specialize in dehalorespiration of lesser chlorinated ethenes produced by PCE-to-DCE dehalorespirers. Maintenance of multiple dehalorespirers growing via complementary substrate interactions results in optimal utilization of electron equivalents, bioenhancement of DNAPL dissolution, and contaminant detoxification.

J. G. Becker (✉)
Department of Civil and Environmental Engineering, Michigan Technological University, Houghton, MI, USA
e-mail: jgbecker@mtu.edu

© Springer Nature Switzerland AG 2020
D. J. O'Bannon (ed.), *Women in Water Quality*, Women in Engineering
and Science, https://doi.org/10.1007/978-3-030-17819-2_9

9.1   Introduction to the Bioremediation of Chlorinated Aliphatic Compounds

9.1.1   Frequency of Contamination and Associated Health Impacts

Chlorinated aliphatic compounds are among the most notorious environmental contaminants, and this group includes 10 of the 20 most common groundwater contaminants at hazardous waste sites [1]. In particular, the chlorinated ethenes tetrachloroethene (PCE) and trichloroethene (TCE) were detected at 60.5% and 51.8%, respectively, of 1300 surveyed contaminated sites [2]. The frequency with which PCE and TCE are detected at contaminated sites is significant because, according to the International Agency for Research on Cancer (IARC), TCE is carcinogenic, and PCE is probably carcinogenic to humans [3]. PCE and TCE are still in use today, primarily as feedstocks in the manufacture of other chemical products, including chlorofluorohydrocarbons and hydrofluorocarbons, as metal degreasers in the automotive and metal industries, and, in the case of PCE, as a dry cleaning solvent.

9.1.2   Processes Affecting the Fate of Chlorinated Hydrocarbons in Groundwater

Figure 9.1 provides an overview of the physicochemical and biological processes that can impact the fate of PCE and TCE in the subsurface. Partitioning of these compounds into different phases occurs after they are introduced to the subsurface because of their physicochemical properties [4]. If sufficient masses of PCE and TCE are present, they may form nonaqueous-phase liquids (NAPLs). NAPL PCE and TCE are denser than water, and the dense NAPLs (DNAPLs) may migrate down into the saturated zone. This is important because dissolution of PCE and TCE from DNAPLs can contaminate groundwater flowing past these pollutant source zones for years. Some PCE and TCE may sorb to the porous medium, but due to their relatively low octanol-water partition coefficient values, most of the contaminant mass remains in the aqueous phase, leading to large contaminant plumes. PCE and TCE are also volatile and thus tend to partition into the gas phase in the unsaturated zone and can migrate in the gas phase. The latter can lead to vapor intrusion and indoor air contamination in overlying buildings.

The chlorinated ethenes are also subject to biological processes in the subsurface [5]. Certain bacteria carry out anaerobic respiration under reduced redox conditions, using chlorinated ethenes as terminal electron acceptors. This process is known as dehalorespiration and involves reductive dechlorination reactions that produce lesser chlorinated ethenes. A number of strains that utilize a range of electron donors

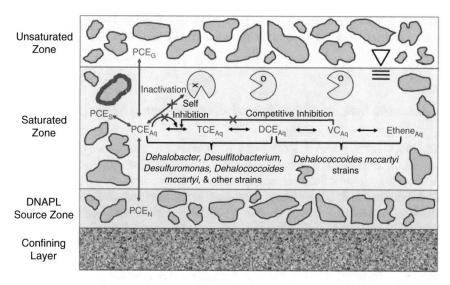

Fig. 9.1 Tetrachloroethene (PCE) partitioning and reductive dechlorination reactions in the subsurface. Subscripts indicate chlorinated ethene phase (Aq = aqueous, N = dense nonaqueous phase liquid, G = gas, S = sorbed). Trichloroethene (TCE), dichloroethene isomers (DCE), and vinyl chloride (VC) also partition into the phases shown

are able to mediate one or more often two dechlorination reaction and produce dichloroethene isomers (DCE) as their dominant product [6]. These bacteria are referred to here as PCE-to-DCE dehalorespirers. Only members of the *Dehalococcoides mccartyi* species have been shown to have the ability to grow via dehalorespiration of DCE and vinyl chloride (VC) and, thus, achieve complete detoxification of PCE or TCE [7]. Reductive dechlorination of PCE and lesser chlorinated ethenes via cometabolic reactions that do not contribute to bacterial growth also occurs but at much slower rates compared to dehalorespiration reactions. Oxidation of *cis*-DCE and VC has been shown to support growth under aerobic conditions, e.g., by *Polaromonas* strains, and cometabolism of chlorinated ethenes with one to three chlorine substituents is feasible [5]. Finally, if chlorinated ethenes are transported to the rhizosphere, they may be taken up by plants and subject to several phytoremediation processes.

In some cases, multiple biological transformation processes may act on chlorinated ethenes [8]. For example, wetland plants transport oxygen to their roots, which creates, within the rhizosphere, aerobic zones that are surrounded by the bulk anaerobic sediment. At sites where groundwater contaminated with PCE and TCE discharges to wetlands, a common scenario because many hazardous waste sites are located close to surface waters, partial transformation of the parent chlorinated ethenes to DCE and/or VC is likely to occur in the anaerobic sediment. A novel rhizosphere bioreactor was developed to evaluate the fate of *cis*-DCE in the wetland rhizosphere in such a scenario [5]. In this system, *cis*-DCE was removed primarily via aerobic oxidation by rhizospheric bacteria and phytovolatilization. However, the focus in this chapter is on the dehalorespiration of chlorinated ethenes because of the importance of this process in the remediation of PCE and TCE at many ground-water sites.

9.1.3   Intrinsic and Engineered Bioremediation Strategies

The in situ (or in place) treatment of subsurface chlorinated ethene contamination can, in some cases, be achieved via intrinsic bioremediation or natural attenuation [8]. Intrinsic bioremediation is feasible only if dehalorespiring populations capable of dechlorinating PCE to ethene are present at the site, and, along with physico-chemical processes, are able to reduce contaminant concentrations to regulatory levels. While PCE-to-DCE dehalorespiring populations are frequently native to chlorinated ethene-contaminated sites, the key *D. mccartyi* strains that dehalore-spire DCE and/or VC are often lacking. Further, the concentrations of electron donors are frequently too low to support complete dechlorination of the chlorinated ethene contaminants. Engineered bioremediation approaches are designed to over-come the limitations on the intrinsic bioremediation capacity at a contaminated site. Therefore, engineered bioremediation approaches implemented at chlorinated ethene-contaminated sites generally include bioaugmentation, i.e., addition of the *D. mccartyi* strains needed to detoxify chlorinated ethenes, and/or biostimulation: the addition of exogenous electron donor(s) needed to sustain complete reductive dechlorination of the contaminants.

9.2   The Importance of Accurate Dehalorespiration Kinetic Models and Parameter Estimates

9.2.1   Kinetic Models of Dehalorespiration

Reliable prediction of the activity of dehalorespiring bacteria and their impacts on contaminant behavior in groundwater systems can only be achieved if accurate and meaningful kinetic parameter estimates are input into these models. Mathematical models of dehalorespiration are generally based on Monod growth kinetics [9], according to:

$$\frac{dX}{dt} = -Y\frac{dS_a}{dt} - bX = \frac{q_{max}YXS_a}{K_{S,a} + S_a} - bX \qquad (9.1)$$

where X is the biomass concentration [$M_X\ L^{-3}$]; t is time [T]; Y is the true yield coefficient [$M_X\ M_S^{-1}$]; S_a is the chlorinated ethene (electron acceptor, a) concentration [$M_S\ L^{-3}$]; b is the first-order decay coefficient [T^{-1}]; q_{max} is the maximum specific substrate utilization rate $[M_S \ M_X^{-1} \ T^{-1}]$; and $K_{S,a}$ is the half-saturation constant for the electron acceptor $[M_S \ L^{-3}]$. At the elevated chlorinated ethene concentrations that might develop in groundwater systems, the Monod model is modified to incorporate appropriate inhibition and/or biomass inactivation terms, as described below.

9.2.2 Obtaining Meaningful and Independent Estimates of Parameter Values

Equation (9.1), and estimates of Y and b obtained in separate assays, must be fit to experimental data to obtain the accurate estimates of q_{max} and K_S needed to model the growth of dehalorespiring populations and chlorinated ethene dehalorespiration. Typically, the experimental data are obtained by monitoring the abundance of chlorinated ethenes over time via headspace analysis in batch assays. When designing batch kinetic assays, three key criteria must be met to ensure that the estimates of q_{max} and K_S are meaningful. These criteria are summarized in Table 9.1.

Proper selection of the initial experimental conditions in the batch assays is the key to achieving the criteria in Table 9.1. For example, the ratio of the initial limiting

Table 9.1 Key criteria for design of batch kinetic assays to ensure meaningful parameter estimates [13]

Batch kinetic assay design consideration	Significance	Quantitative measure of assessing criterion[a]
Estimates must reflect relevant growth conditions, which in turn control microbial community and cellular composition (including the availability of catalytic enzymes)	To estimate kinetic characteristics "extant" in a substrate-limited environment (e.g., a groundwater aquifer), growth must not occur during the assay to preserve the cellular composition; to estimate the "intrinsic" kinetic properties of a culture, unrestricted growth must occur during the assay	Initial substrate-to-biomass (S_0/X_0) ratio calculated on a COD basis
Kinetic parameter estimates must be unique and identifiable	Parameter uniqueness: Unique q_{max} and K_S values can be fit to experimental data independently; nonunique values can be fit to data only as a lumped parameter	Coefficient of determination (R^2)
	Parameter identifiability: Determines amount of correlation; different combinations of highly correlated estimates may fit data equally well	Collinearity index (γ_K)
Parameter estimates should have relatively low uncertainties	Uncertainty cannot be eliminated because experimental data are used to fit unknown kinetic properties; uncertainty in parameter estimates increases with the degree of correlation	Relative standard deviation $(\sigma(\theta)/\theta)$, where θ is a Monod parameter

[a]See text and Reference [13] for guidance on interpreting numerical values of these parameters

substrate concentration (S_0) [M_S L^{-3}] to the initial biomass concentration (X_0) [M_X L^{-3}] determines the growth conditions during the assay. If S_0/X_0 is sufficiently large, the bacterial population will grow without restriction, and the intrinsic kinetic parameter values estimated under these conditions will not be impacted by the culture's history, including any substrate limitations in the source environment [10, 11]. While previous guidelines recommended that intrinsic kinetics be estimated with a $S_0/X_0 > 20$ when both concentrations are expressed on a Chemical Oxygen Demand (COD) basis [10], Huang and Becker [11] showed that estimates of intrinsic q_{max} and K_S for chlorinated ethene respiration by *Desulfuromonas michiganensis* and *Desulfitobacterium* sp. strain PCE1 could reliably be obtained at $S_0/X_0 \geq 10$ (on a COD basis). The finding that intrinsic parameter estimates can be obtained at $S_0/X_0 \geq 10$ is significant because at elevated concentrations, chlorinated ethenes can be inhibitory to, or even inactivate, dehalorespiring bacterial cells. Therefore, it is desirable to use the lowest S_0 possible that will still result in intrinsic parameter estimates for chlorinated ethene respiration. The intrinsic kinetic parameter estimates did not accurately predict the experimental dehalorespiration data collected in assays conducted at lower S_0/X_0 ratios because, under these conditions, the abundance of enzymes that catalyze dehalorespiration reactions is impacted by the substrate availability in the source culture and cell growth is restricted. To completely prevent growth and thus obtain parameters estimates that describe extant microbial activity, i.e., the microbial activity currently in existence in the environment from which the bacteria were obtained, it is recommended that kinetic assays be designed so that the S_0/X_0 ratios are on the order of 0.025 (on a COD basis) [10]. Unfortunately, the initial conditions used in batch kinetic assays are not often reported in the literature and/or are in range (~1 < S_0/X_0 < 10) that will yield neither intrinsic or extant parameter estimates. Parameter estimates obtained at these intermediate S_0/X_0 values cannot be reliably used in mathematical models.

The initial batch assay conditions also affect the uniqueness, identifiability, and hence the amount of uncertainty associated with estimates of q_{max} and K_S [10, 12]. Huang et al. [13] used a combination of numerical experiments and laboratory kinetic assays to systematically and quantitatively evaluate the impact of S_0/X_0 and S_0/K_S on parameter correlation based on the measures in Table 9.1. While S_0/X_0 is key to obtaining intrinsic or extant parameter estimates that reflect the desired physiological state of the bacteria, the numerical and experimental assays indicated that S_0/X_0 does not have a clear impact on parameter correlation. Instead, the study by Huang et al. demonstrated that $S_0/K_S \geq 4$ must be maintained in order to obtain unique and identifiable extant parameter estimates with low uncertainties. In practice, several factors may make it difficult to achieve a S_0 high enough to meet this criterion, particularly when estimating extant kinetics. First, S_0 may be so high when estimating kinetics for PCE and other self-inhibitory compounds that Monod kinetics do not apply, especially if K_S is relatively high. Second, it may be difficult to obtain a S_0 that meets the S_0/K_S ratio while achieving a S_0/X_0 low enough to limit growth so that extant parameter estimates can be obtained. Finally, not knowing the magnitude of K_S *a priori* makes it difficult to choose an appropriate S_0. Therefore, the iterative approach to choosing initial conditions and performing kinetic assays developed by Huang et al. [13] should be adopted to ensure that meaningful extant parameters are estimated. Although this approach was developed and validated using four PCE-respiring strains, it is broadly applicable to parameter estimation for a wide range of substrates and biodegradation processes. The failure to meet the criteria in Table 9.1 when conducting batch kinetic assays undoubtedly contributes to the considerable variability in reported estimates of q_{max} and K_S, even for a given dehalorespiring strain, and consequently, substantially impacts predictions of contaminant fate and transport [13, 14].

9.2.3 Dehalorespiration Kinetic Model that Incorporates Substrate Self-inhibition and Biomass Inactivation Effects

Aqueous chlorinated ethene concentrations are elevated in many groundwater systems impacted by PCE and TCE, particularly those containing NAPL contaminants, and may approach solubility limits. Under these conditions, dehalorespiration rates decrease with increases in the chlorinated ethene concentration and cannot be described using the Monod model (Eq. 9.1). In fact, preliminary dechlorination assays conducted while developing the experimental approach for studies of the relationship between the S_0/X_0 ratios and parameter identifiability [13] showed that when X_0 was ≤ 0.55 mg protein L^{-1}, dechlorination of PCE at concentrations of approximately 700 µM or higher by *D. michiganensis* and *Desulfitobacterium* sp. strain PCE1 ceased completely within several hours [15]. Similar observations were made in another study, in which the Luong model was applied to describe the cessation of dehalorespiration by *D. michiganensis* at elevated concentrations, according to:

$$\frac{dS_a}{dt} = -q_{max} X \frac{S_a}{K_{S,a} + S_a} \left(1 - \frac{S_a}{S_{a-max}} \right)^n \tag{9.2}$$

where S_{a-max} [M$_S$ L^{-3}] is the maximum concentration of PCE at which dehalorespiration proceeds and n is a dimensionless power term, generally assumed to be equal 1, that describes how the specific substrate utilization rate changes in response to increases in elevated S_a concentrations that approach S_{a-max} [16]. However, in assays in which X_0 was several-fold higher than 0.55 mg protein L^{-1}, dehalorespiration proceeded even when the PCE concentration was at the aqueous solubility limit [15]. This indicates that S_{a-max} for PCE is not constant, and therefore, the Luong model is not phenomenologically appropriate or capable of accurately describing PCE dehalorespiration for a broad range of X_0. Further, the log K_{OW} (logarithm of the octanol/water partition coefficient) for PCE at 20 °C (2.6 [4]), is typical of solvents that accumulate in the bacterial cytoplasmic membrane and hence are toxic

[17]. Thus, Huang and Becker [15] used an integrated modeling and experimental approach to test the hypothesis that chlorinated solvents such as PCE can negatively impact dehalorespiration via a self-inhibitory impact on enzymatic reactions plus inactivation of the biomass through solvent effects on membranes. Biomass inactivation is incorporated into the dehalorespiration growth equation according to:

$$\frac{dX}{dt} = -Y\frac{dS_a}{dt_{SI}} - bX - I_B S_a \tag{9.3}$$

where I_B [$M_X M_S^{-1} T^{-1}$] is the biomass inactivation coefficient and dS_a/dt_{SI} is the rate of substrate utilization at self-inhibitory concentrations of PCE. Self-inhibitory utilization of PCE is modeled according to

$$\frac{dS_a}{dt_{SI}} = -q_{max}X\frac{S_a}{K_{S,a} + S_a + \dfrac{S_a^2}{K_{AI}}} \tag{9.4}$$

where K_{AI} [$M_S L^{-3}$] is the self-inhibition constant. The model simulation results obtained using I_B and K_{AI} estimates that were fit to the experimental data demonstrated that at low X_0, biomass inactivation by PCE resulted in negative net growth and prevented dehalorespiration from proceeding (Fig. 9.2). At higher X_0, enough biomass survives to allow dehalorespiration to occur. Conard et al. [18] recently expanded the model to incorporate biomass inactivation by TCE and cis-DCE. The toxicity effects of lesser chlorinated ethenes are the same as that of PCE at a given cell membrane concentration, which is a function of their tendency to partition into the cell membrane from the aqueous phase. An accurate dehalorespiration model that incorporates inhibition and biomass inactivation effects is critical for accurate prediction of contaminant fate, and it can also be used to determine the mass of inoculum needed to sustain dehalorespiration in engineered bioremediation efforts involving bioaugmentation.

9.3 Integrated Modeling and Experimental Evaluation of the Ecology of Chlorinated Hydrocarbon-Degrading Microbial Communities and the Implications for the Implementation of Bioremediation

9.3.1 A Theoretical Modeling Framework of Ecological Interactions Between Dehalorespiring Bacteria

A fundamental tenet of microbial ecology is that interactions between populations within a microbial community are generally based on substrates or other factors utilized for growth. These interactions may be competitive or cooperative.

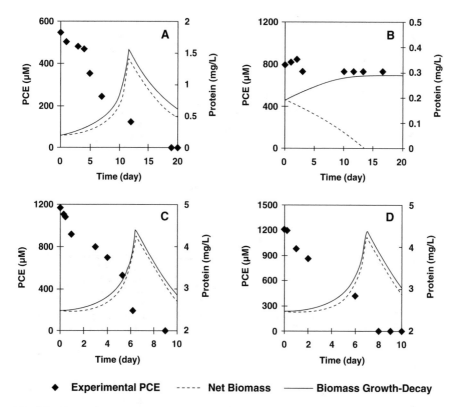

Fig. 9.2 Observed PCE concentrations in *Desulfuromonas michiganensis* cultures containing different initial biomass concentrations (X_0). The net biomass concentration and biomass concentration resulting from growth plus decay simulated using the Andrews biomass inactivation model are also shown. The difference between the two biomass concentrations represents biomass inactivated by exposure to high PCE concentrations ($S_{PCE,0}$). (**a**) $X_0 = 0.19$ mg protein/L, $S_{PCE,0} = 543$ µM; (**b**) $X_0 = 0.19$ mg protein/L, $S_{PCE,0} = 799$ µM; (**c**) $X_0 = 2.5$ mg protein/L, $S_{PCE,0} = 1183$ µM; (**d**) $X_0 = 5.0$ mg protein/L, $S_{PCE,0} = 1183$ µM. (Reprinted with permission from Huang and Becker [15]. Copyright (2011) American Chemical Society)

For example, within chlorinated hydrocarbon-degrading microbial communities, dehalorespiring bacteria often have to compete for hydrogen and/or other electron donors [19]. Becker [6] introduced the idea that when multiple dehalorespiring populations are present in chlorinated ethene-impacted groundwater systems, either as part of the indigenous microbial community or through bioaugmentation (or both), the contaminants themselves may serve as the basis of ecological interactions between dehalorespiring populations.

Three general conceptual models of competitive interactions between dehalorespiring bacteria were formulated based on the substrate ranges of characterized PCE- and TCE-respiring strains and the availability of electron donors during groundwater bioremediation (Fig. 9.3). All three models involve a *D. mccartyi* strain because only members of this hydrogenotrophic species are thought to have

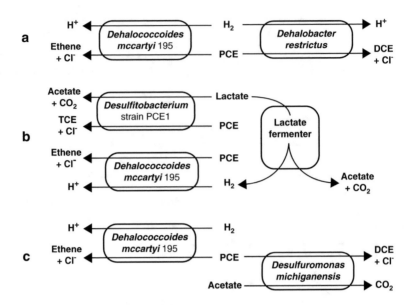

Fig. 9.3 Conceptual models of competition between key dehalorespiring bacteria. (**a**) Scenario 1; (**b**) Scenario 2; and (**c**) Scenario 3. (Adapted with permission from Becker [6]. Copyright (2006) American Chemical Society)

the ability to respire DCE and VC and achieve total detoxification of chlorinated ethenes.

- Scenario 1: *D. mccartyi* competes with another hydrogenotrophic dehalorespirer, (e.g., *Dehalobacter restrictus*) for the electron donor (H_2) and/or common electron acceptors (PCE and TCE; Fig. 9.3a).
- Scenario 2: A fermenter produces acetate and H_2 from lactate, for which it competes with an organotrophic dehalorespirer (e.g., *Desulfitobacterium* sp. strain PCE1). Thus, the organotrophic dehalorespirer and *D. mccartyi* may compete directly for PCE and TCE and indirectly for electron donor equivalents (Fig. 9.3b).
- Scenario 3: An organotrophic dehalorespirer (e.g., *Desulfuromonas michiganensis*) competes with *D. mccartyi* for PCE and TCE, but the two populations use mutually exclusive electron donors (Fig. 9.3c).

Mathematical modeling and literature-derived kinetic parameter estimates were used to simulate the abundance of these populations and contaminant fate in the co-cultures maintained in a continuous-flow stirred tank reactor (CSTR) in the three scenarios defined above [6]. These simulations provide a theoretical framework for understanding and predicting the practical implications of these ecological interactions on the outcome of bioremediation initiatives. The results of this study suggested that under electron donor-limited conditions, which typically prevail at sites undergoing natural attenuation, the faster substrate utilization kinetics of the dehalorespirers competing with *D. mccartyi* allowed them to become dominant, and, in some cases, *D. mccartyi* appeared to be susceptible to washout under these conditions. As a result, PCE was not completely dechlorinated, and TCE or DCE was the predominant product. Importantly, these results demonstrate that, even if *D. mccartyi* is present at a site, complete detoxification of PCE and TCE can be achieved only if *D. mccartyi* can successfully compete with other dehalorespiring populations that are present. Under several scenarios, coexistence of the two deha-lorespiring populations was predicted. In all cases, *D. mccartyi* specialized in grow-ing on the lesser chlorinated ethenes produced by dehalorespiring populations that are capable of only partially dechlorinating PCE to TCE or DCE. An example of this type of complementary interaction is shown in Fig. 9.4a for the two strains in the Scenario 1 co-culture.

9.3.2 Ecological Interactions in a Dehalococcoides-Dehalobacter Co-culture Degrading Aqueous-Phase Contaminant in a Continuous-Flow Stirred-Tank Reactor

Whether the ecological interaction between dehalorespiring populations is of a competitive (e.g., Fig. 9.3a) or complementary (e.g., Fig. 9.4a) nature has poten-tially important implications with respect to the survival of the dehalorespirers and

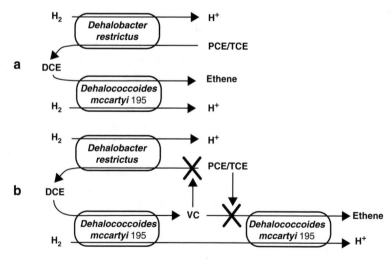

Fig. 9.4 Potential complementary interactions between *D. restrictus* and *D. mccartyi* 195: (**a**) DCE produced by *D. restrictus* is utilized as the electron acceptor by *D. mccartyi* 195 and (**b**) compounded inhibition of dehalorespiration by the electron acceptors and their dechlorination products. (Adapted with permission from Lai and Becker [20]. Copyright (2013) American Chemical Society)

the extent of contaminant detoxification that is feasible in the subsurface. Based on the theoretical framework provided by Becker [6], two hypotheses were formulated. First, competitive interactions will dominate when the availability of the electron donor and/or acceptors is limiting. Second, complementary interactions in which PCE-to-DCE dehalorespirers produce lesser chlorinated ethenes that are utilized as electron acceptors to sustain growth of a *D. mccartyi* strain will be feasible under excess electron donor conditions.

The above hypotheses were experimentally tested by maintaining a co-culture of *D. mccartyi* strain 195 and *D. restrictus* in a CSTR supplied with H_2, acetate, and PCE concentrations that are typical of an engineered bioremediation scenario [20]. When steady state was reached with respect to chlorinated ethene concentrations, PCE was converted primarily to VC, indicating that *D. mccartyi* 195 was able to utilize excess H_2 to respire lesser chlorinated ethenes, as expected. The experimental data were compared to the results predicted using a modified dual Monod model:

$$\frac{dS_a}{dt} = -q_{max} X \left(\frac{S_a}{K_{S,a} + S_a} \right) \left(\frac{S_{donor} - S_{threshold}}{K_{S,donor} + \left(S_{donor} - S_{threshold} \right)} \right) \tag{9.5}$$

where S_{donor} [M_S L^{-3}] is the electron donor concentration, $K_{S,donor}$ [M_S L^{-3}] is the electron donor half-saturation constant, and $S_{threshold}$ [M_S L^{-3}] is the electron donor concentration below which the substrate cannot be utilized [11]. Monod kinetic parameter model inputs were obtained in separate batch assays that were carefully designed following the procedures developed in previous studies. The guidelines of Huang and Becker [11] were used to obtain parameter estimates that describe the intrinsic growth of the dehalorespirers when the limiting substrate concentrations are high relative to the biomass concentrations, conditions that prevailed during the CSTR start-up phase. The procedures of Huang et al. [13] were followed to obtain parameter estimates representing dehalorespirer activity extant in the CSTR following the start-up period and were used to simulate its performance at steady state. While the modified dual Monod model accurately predicted the effluent chlorinated ethene concentrations (Fig. 9.5b), it did not accurately predict the absolute abundance of the dehalorespirers measured using quantitative polymerase chain reaction (qPCR) during steady-state operation [20]. In particular, it substantially overestimated the abundance of *D. restrictus* (Fig. 9.5a).

An additional set of batch kinetic assays were conducted to assess the potential for competitive inhibition of dehalorespiration. In particular, because the CSTR steady-state VC concentration was relatively high (~15 μM), its impact on PCE and TCE respiration by *D. restrictus* was of interest. VC was shown to competitively inhibit PCE and, to an even greater extent, TCE dehalorespiration by *D. restrictus*. In addition, PCE and TCE competitively inhibited VC dechlorination by *D. mccartyi* 195. A conceptual model of the ecological interactions between *D. mccartyi* and *D. restrictus* that includes these effects is shown in Fig. 9.4b. The appropriate competitive inhibitive terms were incorporated in the modified dual Monod model. The resulting competitive inhibition model was used to simulate the CSTR experiment,

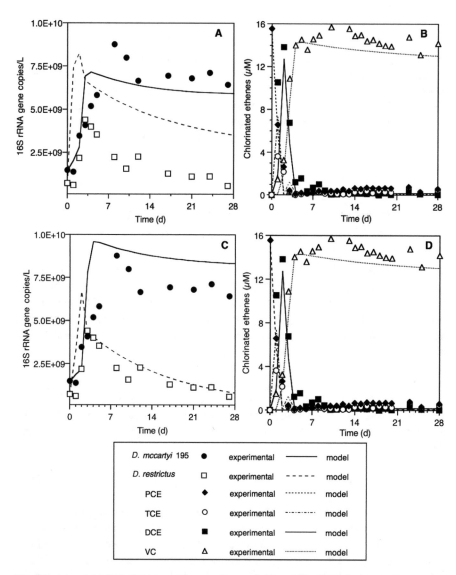

Fig. 9.5 (**a**) and (**c**) 16S rRNA gene copy numbers and (**b**) and (**d**) chlorinated ethene concentrations in a *D. restrictus–D. mccartyi* 195 co-culture growing on PCE and excess H_2 in duplicate CSTRs. (**a**) and (**b**) Lines represent model predictions made using the modified dual Monod dehalorespiration model (Eq. 9.4). (**c**) and (**d**) lines represent model predictions made using the competitive inhibition model. (Adapted with permission from Lai and Becker [20]. Copyright (2013) American Chemical Society)

and showed that the inhibitory effects of PCE and TCE on VC dechlorination contributed to the high concentration of VC in the CSTR, which compounded the negative impacts of VC on dehalorespiration by, and growth of, *D. restrictus*. A much-improved fit was also obtained between the measured dehalorespirer population biomass concentrations and those predicted using the inhibition model (Fig. 9.5c). The CSTR study by Lai and Becker [20] highlights the importance of monitoring not only chlorinated ethene concentrations, which were predicted well by the Monod model, but also the abundance of individual dehalorespiring populations and having meaningful and independent parameter estimates. The latter were essential for understanding the full set of ecological interactions between the hydrogenotrophic dehalorespirers and the implications for their long-term survival and potential for contaminant detoxification in bioremediation applications.

9.3.3 Ecological Interactions in a Dehalococcoides-Desulfuromonas Co-culture Degrading Nonaqueous Liquid Phase Contaminant in Porous Media

PCE and TCE may form DNAPLs when released to the subsurface in sufficient quantities. It is estimated that the majority of the Superfund sites on the National Priority list are impacted by DNAPL contaminants [21]. Removal of these pollutant source zones represents a significant challenge that must be overcome in order to successfully bioremediate aqueous-phase contaminant plumes. Unfortunately, the application of physical and chemical cleanup methods to DNAPL source zones may result in broad dispersal of DNAPL contaminants and thus complicate remediation efforts. The use of dehalorespiring bacteria to enhance dissolution of PCE and TCE from the DNAPL to the aqueous phase, a process known as bioenhanced dissolution, is a promising alternative to other DNAPL remediation methods [22]. The mass transfer rate of DNAPL dissolution [$M_S L^{-3} T^{-1}$] can be described according to:

$$\text{Dissolution rate} = K_1 (C_s - C) \qquad (9.6)$$

where K_1 is the first-order dissolution rate constant [T^{-1}], C_S is the aqueous contaminant concentration [$M_S L^{-3}$] in equilibrium with the DNAPL contaminant concentration at the DNAPL-water interface, and C is the aqueous contaminant concentration [$M_S L^{-3}$] in the bulk groundwater flow. Under abiotic conditions, advection and dispersion decrease C below C_S, which creates a driving force for dissolution. Theoretical, laboratory, and field studies have shown that biodegradation of dissolved contaminants by bacteria can further decrease C, increase the concentration gradient, and thus bioenhance dissolution, as summarized in Ref. [22]. Becker and Seagren [22] recognized that the extent to which dissolution can be bioenhanced is determined by the kinetic characteristics of the dehalorespiring population(s) that grow at the DNAPL-water interface, which in turn will be controlled by the ecological interactions between dehalorespiring populations.

A one-dimensional (1-D) cells-in-series mathematical model and analysis of dimensionless numbers that compare the relative magnitude of mass transfer, mass transport, and biodegradation rates were used to evaluate which factors control these ecological interactions and the amount of dissolution bioenhancement. The 1-D modeling and dimensionless number analyses were applied to a saturated porous media domain containing PCE DNAPL in the form of blobs and a co-culture of *D. mccartyi* and *D. michiganensis*, which may interact competitively (Fig. 9.3a), or in a complementary manner.

The 1-D modeling and dimensionless number analysis showed that electron donor concentrations and hydrodynamics are the key factors determining how much dehalorespiring populations bioenhance dissolution of DNAPL contamination. At a low groundwater velocity ($v_x = 3.3 \times 10^{-4}$ m/h) and low groundwater electron donor concentrations that are typical of a site undergoing intrinsic bioremediation, the potential for bioenhancement is minimal because the dehalorespiring populations are severely limited by the delivery of electron donors.

Bioenhancement of dissolution increased dramatically to 33 times the abiotic dissolution rate at the low v_x and higher electron donor concentrations that are typical of an engineered bioremediation scenario and thus resulted in a biodegradation rate that was high relative to the advection rate. At the high v_x (3.3×10^{-2} m/h), advection greatly decreased the bulk aqueous PCE concentration (C) and, as expected [23], substantially increased the abiotic dissolution rate compared to the low v_x condition [22]. However, the decrease in the availability of aqueous PCE concentration also reduced the rate of PCE dehalorespiration and thus the bioenhancement effect observed at the high v_x was much lower—only 4.8 times the abiotic dissolution rate—than that observed at the low v_x.

Although the 1-D model developed by Becker and Seagren [22] is useful, all 1-D models are subject to several limitations with respect to evaluating the bioenhancement of NAPL dissolution. 1-D models cannot be used to simulate dissolution from a NAPL pool, account for the effects of substrate interactions on the spatial distribution of dehalorespiring populations and the resulting impacts on dissolution bioenhancement, or incorporate heterogeneities in biomass, hydraulic conductivity, and other properties that affect v_x and hence dissolution rates. Therefore, Wesseldyke et al. [24] developed a two-dimensional (2-D) coupled flow-transport model that incorporates the ecological interactions between multiple dehalorespiring bacteria and a PCE DNAPL contaminant source zone. Under high electron donor concentration (biostimulation) and low v_x conditions, the fast substrate utilization kinetics of *D. michiganensis* allowed it to outcompete *D. mccartyi* for PCE, which was the rate-limiting substrate. Consequently, *D. michiganensis* grew along the DNAPL-water interface where its rapid utilization of aqueous PCE substantially bioenhanced dissolution (~10 times the abiotic dissolution rate). *D. mccartyi* had almost no impact on dissolution bioenhancement because it grew in an arc away from the DNAPL-water interface, primarily by utilizing DCE produced by *D. michiganensis*. However, *D. mccartyi* still played a critical role by detoxifying the lesser chlorinated ethenes. In comparison, at the high v_x, more rapid delivery of electron donors allowed *D. mccartyi* to build up biomass, initially by growing on DCE. The increase in *D. mccartyi* biomass increased the rate at which it utilized PCE (Eq. 9.1), and it eventually outcompeted *D. michiganensis* for PCE at the domain inlet and contributed to the dissolution bioenhancement. Over time, *D. mccartyi* biomass clogged the porous medium, reduced the hydraulic conductivity, and deflected flow away from the DNAPL-water interface along which *D. michiganensis* was growing. This decreased the electron donor supply to *D. michiganensis*, causing it to go into net decay mode and *D. mccartyi* to become dominant throughout the model domain.

Many of the predictions of Wesseldyke et al. [24] for the low pore water velocity were confirmed by Klemm et al. [25] in experiments in an intermediate-scale sand tank reactor with a PCE DNAPL pool across the bottom. The tank was inoculated with a co-culture of *D. mccartyi* 195 and *D. michiganensis*. PCE dissolution was enhanced approximately two- to three-fold, primarily by *D. michiganensis*, which was dominant along the DNAPL-water interface. *D. mccartyi* 195 grew primarily on DCE on the periphery of the aqueous contaminant plume that developed downgradient. However, qPCR quantification of the two populations revealed that some biomass inactivation occurred in the areas with the highest PCE concentrations, consistent with the findings of Huang and Becker [15].

9.4   Summary and Conclusions

Contamination of groundwater with PCE and/or TCE is common and poses a serious threat to human health. Bioremediation of chlorinated ethene-contaminated sites is challenging due to the complex ecology of dehalorespiring populations, the partitioning of PCE and TCE into DNAPLs, and site heterogeneities. The conceptual models formulated by Becker [6] provided a framework for a body of work that has integrated mathematical modeling and experimental evaluations to systematically advance our understanding of how ecological interactions between dehalorespirers impact the implementation and performance of bioremediation of these complex sites. These advances would not have been possible without identifying and adhering to the criteria needed to obtain meaningful and identifiable kinetic parameter inputs to the mathematical models and benchmarking model predictions to experimental data. Through this process, the conceptual and mathematical models were revised to account for the biomass inactivation and self-inhibitory effects of PCE on dehalorespirer biomass and activity, respectively; the compounded effects of competitive inhibition of chlorinated ethenes on dehalorespiration processes; and the importance of complementary substrate interactions when multiple dehalorespirers are present. Based on these findings, 2-D mathematical models were developed and compared to experimental data to evaluate how hydrodynamics and electron donor availability affect the outcome of dehalorespirer substrate interactions and the bioenhancement of dissolution of chlorinated ethenes from DNAPL source zones.

Several key themes that have important practical implications have emerged from this body of work. First, PCE-to-DCE dehalorespirers generally have relatively fast substrate utilization kinetics that generally allow them to outcompete *D. mccartyi* strains for higher chlorinated ethenes. Under some conditions, competition may lead to the eventual washout of *D. mccartyi* strains and incomplete detoxification of the contaminant. However, particularly under excess electron donor conditions, maintenance of a *D. mccartyi* strain that specializes in growth on lesser chlorinated ethenes along with a PCE-to-DCE dehalorespirer may be possible. In fact, these studies show that maintaining multiple dehalorespiring populations at PCE-contaminated sites has numerous benefits. In particular, bioaugmentation and biostimulation treatments should be designed to sustain not only *D. mccartyi* strains but also an organotrophic, PCE-to-DCE dehalorespirer such as *D. michiganensis*. This strategy maximizes the fractions of exogenous electron equivalents that are utilized in dehalorespiration, the bioenhancement of DNAPL source zone dissolution and the associated reduction in site cleanup times, and detoxification of the parent contaminants.

Acknowledgments   Professor Becker's research on the microbial ecology and bioremediaton of chlorinated ethene-contaminated sites has primarily been supported by the National Science Foundation through the Presidential Early Career Awards for Scientists and Engineers (PECASE) award that she received under Grant No. 0134433 and through Grant No. 1034700, which was awarded to Professors Becker and Eric A. Seagren (Michigan Tech).

References

1. AFCEE (2000) Remediation of chlorinated solvent contamination on industrial and airfield sites. Air Force Center for Environmental Excellence Brooks Air Force Base, TX
2. USEPA (1991) Revised priority list of hazardous substances. Fed Regist 56(201): 521654–552175
3. IARC (2012) Trichloroethylene, tetrachloroethylene, and some other chlorinated agents. In IARC monographs on the evaluation of carcinogenic risks to humans. International Agency for Research on Cancer (IARC) working group on the evaluation of carcinogenic risks to humans, vol 106, Lyon, France
4. Versheuren K (1983) Handbook of environmental data on organic chemicals, 2nd edn. Van Nostrand Reinhold Co, New York
5. Tawney I, Becker JG, Baldwin AH (2008) A novel dual compartment, continuous-flow wetland microcosm to assess *cis*-dichloroethene removal from the rhizosphere. Int J Phytoremediation 10:455–471
6. Becker JG (2006) A modeling study and implications of competition between *Dehalococcoides ethenogenes* and other tetrachloroethene-respiring bacteria. Environ Sci Technol 40:4473–4480
7. Löffler FE et al (2013) *Dehalococcoides mccartyi* gen. nov., sp. nov., obligate organohalide-respiring anaerobic bacteria, relevant to halogen cycling and bioremediation, belong to a novel class, *Dehalococcoidetes classis* nov., within the phylum Chloroflexi. IJSEM 63:625–635
8. Becker JG, Seagren EA (2009) Bioremediation of hazardous organics. In: Mitchell R, Gu J-D (eds) *Environmental microbiology*, 2nd edn. Wiley-Blackwell, Hoboken, pp 177–212
9. Monod J (1949) The growth of bacterial cultures. Ann Rev Microbiol 3:371–394
10. Grady CPL Jr, Smets BF, Barbeau DS (1996) Variability in kinetic parameter estimates: a review of possible causes and a proposed terminology. Water Res 30(3):742–748
11. Huang D, Becker JG (2009) Determination of intrinsic Monod kinetic parameters for two heterotrophic tetrachloroethene (PCE)-respiring strains and insight into their application. Biotechnol Bioeng 104(2):301–311

12. Liu C, Zachara JM (2001) Uncertainties of Monod kinetic parameters nonlinearly estimated from batch experiments. Environ Sci Technol 35(1):133–141
13. Huang D, Lai Y, Becker J (2014) Impact of initial conditions on extant microbial kinetic parameter estimates: application to chlorinated ethene dehalorespiration. Appl Microbiol Biotechnol 98(5):2279–2288
14. Chambon JC et al (2013) Review of reactive kinetic models describing reductive dechlorination of chlorinated ethenes in soil and groundwater. Biotechnol Bioeng 110(1):1–23
15. Huang D, Becker JG (2011) Dehalorespiration model that incorporates the self-inhibition and biomass inactivation effects of high tetrachloroethene concentrations. Environ Sci Technol 45(3):1093–1099
16. Amos BK et al (2007) Experimental evaluation and mathematical modeling of microbially enhanced tetrachloroethene (PCE) dissolution. Environ Sci Technol 41:963–970
17. Fernandes P, Ferreira BS, Cabral JMS (2003) Solvent tolerance in bacteria: role of efflux pumps and cross-resistance with antibiotics. Int J Antimicrob Agents 22:211–216
18. Conard S (2016) Modeling Bioenhanced bioenhanced DNAPL Pool pool Dissolutiondissolution: Sensitivity sensitivity Analysisanalysis, Inhibition inhibition Kinetic kinetic Effectseffects, and Intermediateintermediate-Scale scale Flow flow Cell cell Experiment experiment Evaluationevaluation. M.S. Thesisthesis. Michigan Technological University, Houghton, MI
19. Fennell DE, Gossett JM (1998) Modeling the production of and competition for hydrogen in a dechlorinating culture. Environ Sci Technol 32:2450–2460
20. Lai Y, Becker JG (2013) Compounded effects of chlorinated ethene inhibition on ecological interactions and population abundance in a *Dehalococcoides-Dehalobacter* co-culture. Environ Sci Technol 47:1518–1525
21. USEPA (1993) Evaluation of the likelihood of DNAPL presence at NPL sites, national results. Office of Solid Waste and Emergency Response (OSWER)
22. Becker JG, Seagren EA (2009) Modeling the effects of microbial competition and hydrodynamics on the dissolution and detoxification of dense nonaqueous phase liquid contaminants. Environ Sci Technol 43:870–877
23. Seagren EA, Moore TO II (2003) Nonaqueous phase liquid pool dissolution as a function of average poor water velocity. J Environ Eng 129:786–799
24. Wesseldyke ES et al (2015) Numerical modeling analysis of hydrodynamic and microbial controls on DNAPL pool dissolution and detoxification: dehalorespirers in co-culture. Adv Water Resour 78:112–125
25. Klemm, S (2016). Bioenhanced Dissolution dissolution of a Tetrachloroethene tetrachloroethene (PCE) Pool pool in a Sand sand Tank tank Aquifer aquifer Systemsystem. M.S. Thesisthesis. Michigan Technological University, Houghton, MI

Jennifer G. Becker is an Associate Professor of Environmental Engineering at her alma mater, Michigan Technological University. Michigan Tech is located in the beautiful Upper Peninsula of Michigan and is just a few miles from Lake Superior, which is constant source of inspiration for those working to protect water quality. Previously, Professor Becker was an Associate Professor and Extension Specialist at the University of Maryland, College Park.

Professor Becker saw firsthand the extensive tree damage from acid rain in Germany's Black Forest as a girl, which convinced her to pursue an environmental engineering degree so that she could apply her STEM skills to the solution of challenging environmental problems.

Dr. Becker worked for Zimpro/Passavant (now part of Siemens) designing wastewater treatment facilities after her B.S. Environmental Engineering degree from Michigan Tech. She then went on to pursue a M.S. degree in environmental engineering at the University of Illinois at Urbana. Her M.S. thesis advisor, Dr. David L. Freedman, introduced her to the field of bioremediation, the use of microorganisms to detoxify pollutants and cleanup contaminated sites. Bioremediation of groundwater contaminated with chlorinated solvents continues to be an important focus of Professor Becker's research in the area of biological treatment processes. Through her work with David Freedman, she acquired expertise in a wide range of microbiological and analytical methods, an appreciation of the importance of scientific rigor in conducting research, effective technical writing skills, and, most importantly, a passion for research. Professor Becker went on to pursue a Ph.D. degree in environmental engineering under the direction of Drs. Bruce E. Rittmann and David A. Stahl at Northwestern University. Her Ph.D. studies gave her expertise in molecular ecology and environmental biotechnology and an interest in understanding how the ecology of microbial communities fundamentally influences the fate of environmental contaminants.

Professor Becker received a 2002 National Science Foundation Early Career Development (CAREER) Award in support of her work on groundwater bioremediation. The innovative nature and theoretical and practical importance of her work was also recognized with a Presidential Early Career Award for Scientists and Engineers (PECASE) given to Professor Becker at a White House ceremony in 2002. The PECASE is the highest honor bestowed by the US government on outstanding scientists and engineers beginning their independent careers. Professor Becker was the recipient of the Association of Environmental Engineering and Science Professors (AEESP)/ Montgomery Watson Harza Master's Thesis Award (both as faculty advisor and as a student), the Water Environment Federation Robert A. Canham Award, and several other national and regional awards. She served on the AEESP Board of Directors from 2010 to 2014 and is a Past President of AEESP. Professor Becker served on the Environmental Engineering Committee of the US Environmental Protection Agency (EPA) Science Advisory Board from 2016 to 2018.

Professor Becker derives a great deal of energy and inspiration from her interactions with students, particularly in the classroom. She is passionate about her research projects and the subjects she teaches and finds it gratifying when students share that passion. It is her goal to inspire more students to become educated and trained as environmental process engineers so that they can take on the challenges of addressing today's urgent environmental problems. She is especially interested in doing work that positively and directly impacts the practice of engineering. Professor Becker is married to Professor Eric A. Seagren (also of the Civil and Environmental Engineering Department at Michigan Tech), and they have two children. Her parents continue to be model environmental stewards, and from them, she acquired not only her love of nature but also, from their example, learned how to be resilient and self-assured.

Chapter 10
Fate of Veterinary Pharmaceuticals in Agroecosystems

Shannon L. Bartelt-Hunt

Abstract Veterinary pharmaceuticals, which are increasingly used in animal production practices, can enter surface and groundwater after land application of animal manures or animal wastewater. The presence of veterinary pharmaceuticals can result in negative environmental impacts including the proliferation of environmental antibiotic resistance and endocrine-disrupting effects in aquatic organisms. The efficacy of manure application strategies to limit the occurrence of veterinary pharmaceuticals in runoff and best management practices to remove these compounds from runoff prior to entering surface water should be investigated to mitigate the impact of these compounds on the environment.

10.1 Introduction

Veterinary pharmaceuticals are used regularly in the livestock industry as growth promoters, to improve feed efficiency, for disease prevention, or as part of therapeutic treatment. Biologically active pharmaceuticals used in animal production include antimicrobials, steroid hormones, and beta agonists, such as ractopamine which is used in swine and cattle production. The amount of antimicrobials used in the agriculture industry in the United States has been estimated between 8.5 million kg [31] and 12.6 million kg. Although reliable data regarding the usage of antibiotics in animal production are difficult to find, over half of the antibiotics consumed in the United States are used in animal agriculture [27]. Similarly, steroid hormones are given to nearly all of the approximately 32 million beef cattle produced in the United States annually, with dosage amounts up to hundreds of milligrams, depending on the implant or feed additive administered [24].

S. L. Bartelt-Hunt (✉)
Department of Civil Engineering, University of Nebraska-Lincoln, Lincoln, NE, USA
e-mail: sbartelt2@unl.edu

© Springer Nature Switzerland AG 2020
D. J. O'Bannon (ed.), *Women in Water Quality*, Women in Engineering and Science, https://doi.org/10.1007/978-3-030-17819-2_10

Veterinary pharmaceuticals administered to animals are excreted in manures, and up to 90% of certain pharmaceuticals may be excreted unmetabolized [18]. Pharmaceutical concentrations in manure have been reported to range from trace levels to hundreds of milligrams per kilogram [35, 37]. For centuries, minimally treated animal manures have been applied to agricultural fields as a soil conditioner and fertilizer. In 2015, the United States produced a record high 94 billion pounds of red meat and poultry, despite an 80% decline in the number of animal production facilities since the 1950s [14]. This trend of increasing geographic density of animal production in the United States unavoidably results in increased water quality degradation by conventional contaminants such as nutrients and pathogens as well as veterinary pharmaceuticals due to runoff from fields where animal manures are land applied [40]. The presence and activity of antimicrobials in manure can increase antimicrobial resistant bacteria, even at low antimicrobial concentrations [15], while the occurrence of other veterinary pharmaceuticals, such as steroid hormones, can lead to endocrine-disrupting effects in aquatic organisms [1, 34, 42].

The fate pathways for veterinary pharmaceuticals in agroecosystems are shown in Fig. 10.1. Veterinary pharmaceuticals are released into animal manure or wastewater, which is typically stored on site in manure pits (swine), in stockpiles or compost piles (beef cattle, dairy, and poultry), or in wastewater lagoons. Both animal manures and wastewaters are routinely land applied as an organic fertilizer and soil conditioner. Once applied to land, the veterinary pharmaceuticals can be transported to surface water via runoff or infiltrate into soil and be transported to groundwater. This chapter reviews the occurrence of veterinary pharmaceuticals in surface water and describes practices that may limit their transport after land application of manure to crop fields.

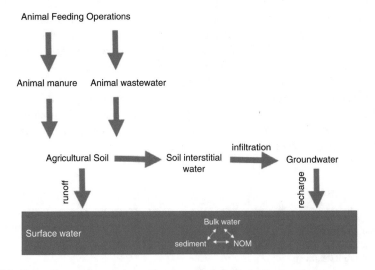

Fig. 10.1 Fate pathways for veterinary pharmaceuticals in the agro-ecosystem

10.2 Occurrence of Veterinary Pharmaceuticals in Agroecosystems

The occurrence of veterinary pharmaceuticals in animal manures has been well-documented. Comprehensive reviews regarding the concentrations and types of veterinary pharmaceuticals in animal manures are provided by Sarmah et al. [35] and Song and Guo [37]. Numerous classes of antimicrobials have been detected in manures from swine, cattle, and poultry production at concentrations ranging from trace levels to hundreds of mg/kg [37]. Although there is significant evidence showing that pharmaceuticals can be transported to surface water in runoff from land-applied manure, to date, a limited number of studies have evaluated the fate and persistence of antibiotics in surface waters in intensively agricultural watersheds with minimal municipal wastewater inputs.

Jaimes Correa et al. [17] previously documented the occurrence and persistence of pharmaceuticals in an intensively agricultural watershed in. In this study, the occurrence of pharmaceuticals was monitored in the Shell Creek watershed in east-central Nebraska. This watershed is approximately 1200 km^2, and the five communities within the watershed have a combined population of 1675 people. By contrast, the counties comprising the watershed include 1550 farms with over one million head of swine, cattle, and poultry. Cultivated land cover within the watershed is 78.2%, while urban developed areas are only 4.4%. During this monitoring study, occurring from September 2008 through October 2009, the presence of 12 veterinary pharmaceuticals was detected using a LC-MS/MS analysis method in at least one sampling event with concentrations ranging from 0.0003 to 68 ng/L (Fig. 10.2). As shown in Fig. 10.2, ANOVA reveals significant differences in mean concentrations between antibiotics ($p < 0.01$). Results from Tukey's multiple comparison test are represented by letters. Antibiotics with similar letters (e.g., "a" and "ab") have no significant differences in mean concentrations ($p > 0.05$) while antibiotics with different letters ("a" and "b") have significant differences in mean concentrations ($p < 0.05$). The compounds detected at the highest time-weighted average (TWA) concentrations in Shell Creek were lincomycin (68 ng/L) and monensin (49 ng/L). Tiamulin, sulfadimethoxine, and sulfamethazine had maximum concentrations of 2.6, 3.9, and 13 ng/L, respectively. Dissolved concentrations of the beta agonist, ractopamine, three sulfonamide-group antibiotics sulfachloropyridazine, sulfamethazole, sulfamethoxazole, and the macrolide tylosin were all detected at average concentrations less than 1 ng/L [17]. In this study, increased antibiotic concentrations were identified in the summer months and were likely driven by rainfall-runoff events [17]. This finding is consistent with other studies that have identified increases in antibiotic concentrations in agricultural watersheds in the summer months [30]. Although some temporal trends were observed, it should be noted that antibiotics were detected in each monthly sampling event, indicating that pharmaceuticals can persist in surface water, even if they are introduced via episodic runoff events. In urban or suburban watersheds, the predominant source of veterinary pharmaceuticals is from municipal wastewater effluents, which are more continuous sources. The

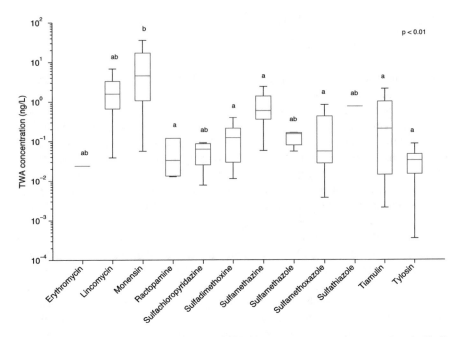

Fig. 10.2 Distribution of time-weighted average (TWA) pharmaceutical concentrations in Shell Creek from Jaimes Correa et al. [17]. The split box shows the 25th, 50th, and 75th, whereas whiskers shows the 5th and 95th percentiles

results indicate that although agricultural ecosystems are less likely to contain significant veterinary pharmaceutical loadings from municipal wastewater, the occurrence of pharmaceuticals in surface waters within these watersheds is persistent.

10.3   Influence of Manure Handling Practices

Cattle and swine manure accumulates within the production facility during the animal production period. In cattle feedlots, manure accumulates within the animal pen and then is typically scraped out at the end of the animal production period, prior to the introduction of new animals. This manure is typically stockpiled for a period of months prior to land application onto crop fields. In swine production systems, typically one of three waste handling systems is used: flush systems, pit recharge, or deep pits [35]. In deep pit systems, manure falls from a slatted floor into a pit below the animal housing facility and typically uses less water than either flush or pit recharge systems [35]. Manure may be stored in these pits for up to a year. Deep pit systems are commonly used in colder climates such as the upper Midwest in the United States, and manure accumulating in deep pits provides an environment for anaerobic microbial activities.

The fate of antimicrobials during anaerobic swine manure storage was evaluated in a previous study [20]. In this study, manure was obtained from an operating swine production facility that contained chlortetracycline, tylosin, and bacitracin A. After collection, manure and water were mixed in a 2:1 (w/w) ratio in 100 mL amber glass reactors, sparged with nitrogen, and incubated at 37 °C for up to 40 days to monitor the persistence of the antimicrobials. The parent antimicrobials tylosin and chlortetracycline were detected in swine manure reactors at initial concentrations of 10 mg/kg (dry weight basis) and 300 mg/kg (dry weight basis), respectively [20]. Bacitracin A was not detected in the manure at any time, but bacitracin F, a metabolite of bacitracin A, was detected at an initial concentration of 50 mg/kg (dry weight basis) in the manure.

Observed antimicrobial concentrations were fit with a first-order reaction equation to determine rate constants and first-order half-lives (Table 10.1). The first-order reaction rate constant for tylosin, chlortetracycline, and bacitracin F were -0.07 d^{-1} ($R^2 = 0.34$), -0.6 d^{-1} ($R^2 = 0.79$), and -0.36 d^{-1} ($R^2 = 0.94$), respectively. The half-life for chlortetracycline measured in Joy et al. [20] is shorter than that reported previously in studies of chlortetracycline degradation in swine manure or soil [6, 28, 39]. In contrast, the tylosin half-life reported in Joy et al. [20] is consistent with previous studies that measured tylosin half-lives on the order of 4.4 days [6, 25].

In Joy et al. [20], the occurrence of the antimicrobials and their corresponding antibiotic resistance genes (ARGs) were monitored. Although the antibiotic concentrations at the end of the 40 day experiments were ~10% of the initial concentration, the relative abundance of certain ARGs were more persistent, with approximately 50% of the initial abundance at the end of the storage period. The differences in observed behavior between the antimicrobials and corresponding ARGs indicates the importance of identifying not only the occurrence of the parent antimicrobial but also any biologically active degradation products, which could continue to exert a selective pressure allowing for the observed proliferation of resistance genes in manure storage systems.

Cattle manure and poultry litter handling have also been evaluated to determine the influence of practices such as composting or stockpiling on veterinary pharmaceutical concentrations. A number of studies have demonstrated the efficacy of composting for reducing the concentrations of nutrients and veterinary pharmaceuticals such as antibiotics and steroid hormones [2–4, 8, 22, 33].

Table 10.1 Antimicrobial degradation rates in simulated swine manure storage from Joy et al. [20]

Antimicrobial	Measured degradation rate (d^{-1})	Measured half-life (d)	Reported half-life (d)
Chlortetracycline	-0.6	1	20–70 d [6, 39]
Tylosin	-0.07	9.7	0.02–4.4 [25, 6]
Bacitracin F	-0.36	1.9	Not available

10.4   Practices to Control Veterinary Pharmaceutical Transport After Land Application of Manures

Contaminants present in the manure including nutrients, pathogens, and trace compounds such as veterinary pharmaceuticals can be transported to surface water following land application. Although the concentrations of conventional pollutants in runoff from crops fertilized with animal manures have been routinely documented, there are few studies investigating the fate and transport of antimicrobials in soil and in runoff following land application of manure. Once animal manure is land applied, the fate of manure-borne compounds in soil and subsequent transport in runoff will be affected by the compounds' sorption properties [7, 23, 36] and susceptibility to biotic and abiotic degradation process such as photolysis [11, 16, 41].

Several studies have investigated the influence of manure application strategy on antimicrobial concentrations in runoff. One study found no statistically significant differences in concentrations of chlortetracycline, monensin, and tylosin in infiltration water and surface runoff when manure was applied using two different land application methods [9]. In contrast, other studies suggest that soil tillage leads to reduced vertical transport of antimicrobials after broadcast application of liquid manure [21], and manure incorporation (i.e., mixing manure into the top soil) could lead to reduced antimicrobial concentrations in runoff [26]. Joy et al. [19] published a study evaluating the influence of manure application methods on the concentration of antimicrobials in soil and runoff after land application of swine manure. In this study, a rainfall simulation study was conducted using test plots (0.75 m × 2.0 m) where swine manure was land applied using one of three land application methods: broadcast, incorporation, or injection. The plots were established using a randomized block design, and on each plot, three sequential rainfall simulation experiments were performed. Control plots with no manure amendment were also subjected to rainfall simulation experiments.

Broadcast manure generally resulted in higher antimicrobial concentrations in runoff than did incorporated and injected manures (Fig. 10.3). Because swine slurry

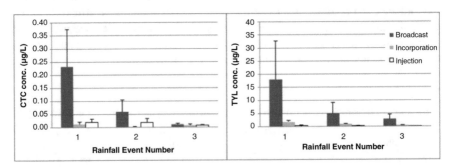

Fig. 10.3 Aqueous concentrations of chlortetracycline (CTC) and tylosin (TYL) in runoff from manure-amended plots receiving broadcast, incorporation, and injection treatments over three rainfall events. Error bars show the standard errors over triplicate field experiments. (Figure reprinted with permission from Joy et al. [19])

was spread on the soil surface in the broadcast application, the antimicrobials were readily available for transport to runoff during rainfall events. In contrast, mixing manure slurry with surface soil to various extents (i.e., injection and incorporation) resulted in reduced transport of antimicrobials to runoff. Although the main treatment factor, application method, was not considered statistically significant according to the rANOVA tests ($p = 0.26$ for chlortetracycline and $p = 0.31$ for tylosin), this is likely due to large variation in observed concentrations among the triplicate plots, which are not uncommon in field-scale experiments. The differences in sorption partition coefficients between chlortetracycline and tylosin to soil might account for the differences in runoff concentrations. It was not surprising to observe that the aqueous chlortetracycline concentrations in the runoff were low (Fig. 10.3) because of its sorptive nature (log K_{ow} for chlortetracycline is −0.62). By contrast, the range of tylosin concentrations in runoff measured in this study was 0.087–18 µg/L (log K_{ow} for tylosin is 1.63).

In addition to manure application practices, other best management practices that have been used historically to control the movement of conventional contaminants such as nutrients or pathogens can also be used to mitigate veterinary pharmaceutical transport, although this remains an underinvestigated research area. Soni et al. [38] investigated the use of narrow grass hedges, a type of vegetative barrier, in controlling antimicrobial runoff from plots amended with swine manure.

Vegetative barriers (VBs) are strips of densely growing plants used primarily on croplands adjacent to surface water. Vegetative barriers can reduce both dissolved and sediment-bound compounds in runoff by reducing runoff volume and capturing sediment [29]. VBs reduce the kinetic energy of the runoff, which can lead to enhanced settling of particulate contaminants. Dissolved contaminants can be reduced by enhanced infiltration and improved water-holding capacity of the surface soil within VBs [29].

Narrow grass hedges (NGH) are one type of vegetative barrier and are constructed using stiff stemmed grass strips that are ~1.5 m wide. Narrow grass hedges have been demonstrated to be effective in removing both dissolved [10, 13, 32] and sediment-bound [12] nutrients from runoff. The potential efficacy of narrow grass hedges for removal of antimicrobials from runoff was evaluated [38]. Similar to as in Joy et al. [19], test plots were established which were amended with swine manure and were established in a randomized block design. Three treatment factors were tested for their effects on runoff water quality: manure amendment (manure application to meet zero vs. three times annual N demand by corn or control vs. amended plots), NGH (plots with and without a NGH), and rainfall events (day 1, 2, and 3). In this set of field experiments, the only antimicrobial measured in the manure that was land applied was tylosin.

ANOVA analysis indicates that both manure amendment and the presence of a NGH had significant effects on the presence of tylosin in runoff ($p < 0.0001$). Although tylosin concentrations in runoff decreased with successive rainfall events (Fig. 10.4), the impacts of this treatment factor were not statistically significant. Prior to this study, little was known about the effectiveness of NGHs on reducing

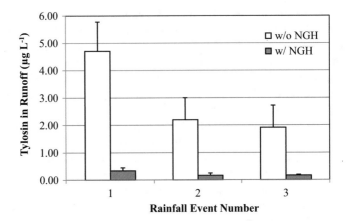

Fig. 10.4 Concentrations of dissolved tylosin in runoff from manure amended plots with and without NGH. Error bars represent standard errors from triplicate field experiments. (Reprinted with permission from Soni et al. [38])

Table 10.2 Mass loadings of tylosin in runoff from the amended plots with and without NGH during three rainfall events (average ± standard error). Averages and standard errors were calculated based on triplicate field experiments. Reprinted with permission from Soni et al. [38]

	Tylosin	
Rainfall event	w/o NGH ($\mu g\ m^{-2}$)	w/ NGH ($\mu g\ m^{-2}$)
1	48.5 ± 23.3	2.74 ± 1.77
2	33.7 ± 13.4	3.61 ± 3.29
3	20.5 ± 12.6	2.48 ± 0.59
Sum	103	8.87
Fraction from #1	0.47	0.31

dissolved antimicrobial loadings and concentrations in runoff. In this study, NGHs lowered tylosin loadings in runoff by more than an order of magnitude (Table 10.2). Another study also demonstrated that vegetative buffer strips made of tall fescue could reduce tylosin in runoff [29]. Enhanced infiltration or adsorption of tylosin within the NGH system likely accounted for increased removal of dissolved tylosin loadings in runoff. The dissolved tylosin concentrations in runoff decreased with successive rainfall events for plots without a NGH, whereas no such trend was observed for plots with a NGH (Fig. 10.4 and Table 10.2). As a cost-effective best management practice, NGHs have been demonstrated to be effective in reducing contaminant loads in agricultural runoff, and the results from Soni et al. [38] also demonstrate that NGH can reduce dissolved antimicrobials in agricultural runoff following land application of swine manure.

10.5 Conclusions

Antibiotics and steroid hormones are regularly used in animal production and are excreted in animal manures. Although land application of manure provides benefits in agricultural production, including the reduction in use of commercial fertilizers, trace organics contained in manure can run off from cropland and contaminate surface and groundwater. The results presented here quantify antibiotic loading in runoff from cropland amended with manure and in surface water within watersheds with significant animal and crop production facilities. It is important to understand the impact of manure management practices on limiting antimicrobial impacts to surface and groundwater. Management practices such as manure storage and composting, manure incorporation into soil during land application, and the use of vegetated buffer strips can all reduce the loading of antibiotics and steroid hormones to the environment; however, more research is needed to evaluate the transformation of trace organics in agricultural production systems, as well as the relationship between the occurrence of antibiotics and antibiotic resistance genes, which can lead to the proliferation of environmental antibiotic resistance.

Acknowledgments I would like to acknowledge my collaborators in the work presented in this chapter including Dr. Xu Li, Dr. Daniel Snow of the University of Nebraska-Lincoln, and Dr. John Gilley of the United States Department of Agriculture. I would also like to thank the many students who worked on these projects including Juan Jaimes Correa, Stacey (Joy) Thomas, and Bhavneet Soni.

References

1. Ali J, D'Souza D, Schwarz K, Allmon L, Singh RP, Snow DD, Bartelt-Hunt SL, Kolok A (2018) Endocrine effects following early life exposure to water and sediment found within agricultural runoff from the Elkhorn River, Nebraska, USA. Sci Total Environ 618:1371–1381
2. Arikan OA, Mulbry W, Rice C (2009) Management of antibiotic residues from agricultural sources: use of composting to reduce chlortetracycline residues in beef manure from treated animals. J Hazard Mater 164:483–489
3. Bao Y, Zhou W, Guan L, Wang Y (2009) Depletion of chlortetracycline during composting of ages and spiked manures. Waste Manag 29:1416–1423
4. Bartelt-Hunt SL, Devivo S, Johnson L, Snow DD, Kranz WL, Mader TL, Shapiro CA, vanDonk SJ, Shelton DP, Tarkalson DD, Zhang TC (2013) Effect of composting on the fate of steroids in beef cattle manure. J Environ Qual 42:1159–1166
5. Carlson JC, Mabury SA (2006) Dissipation kinetics and mobility of chlortetracycine, tylosin and monensin in an agricultural soil in Northumberland County, Ontario, Canada. Environ Toxicol Chem 25(1):10
6. Davis JG, Truman CC, Kim SC, Ascough JC, Carlson K (2006) Antibiotic transport via runoff and soil loss. J Environ Qual 35(6):2250–2260
7. Derby NE, Hakk H, Casey FXM, DeSutter TM (2011) Effects of composting swine manure on nutrients and estrogens. Soil Sci 176:91–98
8. Dolliver H, Gupta S (2008) Antibiotic losses in leaching and surface runoff from manure-amended agricultural land. J Environ Qual 37(3):1227–1237

9. Eghball B, Gilley JE, Kramer LA, Moorman TB (2000) Narrow grass hedge effects on phosphorus and nitrogen in runoff following manure and fertilizer application. J Water Soil Conserv 55:172–176

10. Eichhorn P, Aga DS (2004) Identification of a photooxygenation product of chlortetracycline in hog lagoons using LC/ESI-ion trap-MS and LC/ESI-time-of-flight-MS. Anal Chem 76(20):6002–6011

11. Gilley JE, Eghball B, Marx DB (2008) Narrow grass hedge effects on nutrient transport following compost application. Trans ASABE 51:997–1005

12. Gilley JE, Durso LM, Eigenberg RA, Marx DB, Woodbury BL (2011) Narrow grass hedge control of nutrient loads following variable manure applications. Trans ASABE 54:847–855

13. Graham JP, Nachman KE (2010) Managing waste from confined animal feeding operations in the United States: the need for sanitary reform. J Water Health 8(4):646–670

14. Heuer H, Schmitt H, Smalla K (2011) Antibiotic resistance gene spread due to manure application on agricultural fields. Curr Opin Microbiol 14(3):236–243

15. Hu D, Coats JR (2007) Aerobic degradation and photolysis of tylosin in water and soil. Environ Toxicol Chem 26(5):884–889

16. Jaimes-Correa JC, Snow DD, Bartelt-Hunt SL (2015) Seasonal occurrence of antibiotics and beta agonists in an agriculturally-intensive watershed. Environ Pollut 205:87–96

17. Jjemba PK (2006) Excretion and ecotoxicity of pharmaceutical and personal care products in the environment. Ecotoxicol Environ Saf 63(1):113–130. https://doi.org/10.1016/j.ecoenv.2004.11.011

18. Joy SR, Bartelt-Hunt SL, Snow DD, Gilley J, Woodbury B, Parker D, Marx D, Li X (2013) Fate and transport of antimicrobials and antimicrobial resistance genes in soil and runoff following land application of swine manure slurry. Environ Sci Technol 47(21):12081–12088

19. Joy SR, Li X, Snow DD, Gilley JE, Woodbury B, Bartelt-Hunt SL (2014) Fate of antimicrobials and antimicrobial resistance genes in simulated swine manure storage. Sci Total Environ 481:69–74

20. Kay P, Blackwell PA, Boxall ABA (2004) Fate of veterinary antibiotics in a macroporous tile drained clay soil. Environ Toxicol Chem 23(5):1136–1144

21. Kim KR, Owens G, Ok YS, Park WK, Lee DB, Kwon SI (2012) Decline in extractable antibiotics in manure-based composts during composting. Waste Manag 32:110–116

22. Kim S-C, Davis JG, Truman CC, Ascough JC II, Carlson K (2010) Simulated rainfall study for transport of veterinary antibiotics – mass balance analysis. J Hazard Mater 175(1–3):836–843

23. Kolok AS, Sellin MK (2008) The environmental impact of growth-promoting compounds employed by the United States beef cattle industry: history, current knowledge and future directions. In: Whitacre DM (ed) Reviews of environmental contamination and toxicology

24. Kolz AC, Moorman TB, Ong SK, Scoggin KD, Douglass EA (2005) Degradation and metabolite production of tylosin in anaerobic and aerobic swine manure lagoons. Water Environ Res 44:49–56

25. Kreuzig R, Holtge S, Brunotte J, Berenzen N, Wogram J, Schulz R (2005) Test-plot studies on runoff of sulfonamides from manured soils after sprinkler irrigation. Environ Toxicol Chem 24(4):777–781

26. Landers TF, Cohen B, Wittum TE, Larson EL (2012) A review of antibiotic use in food animals: perspective, policy and potential. Public Health Rep 127:4–22

27. Li L, Huang L, Chung R, Fok K, Zhang Y (2010) Sorption and dissipation of tetracyclines in soils and compost. Pedosphere 20:807–816

28. Lin CH, Lerch RN, Goyne KW, Garrett HE (2011) Reducing herbicides and veterinary antibiotics losses from Agroecosystems using vegetative buffers. J Environ Qual 40:791–799

29. Lissemore L, Hao C, Yang P, Sibley PK, Mabury S, Solomon KR (2006) An exposure assessment for selected pharmaceuticals within a watershed in Southern Ontario. Chemosphere 64(5):717–729

30. Nawaz MS, Erickson BD, Khan AA, Khan SA, Pothuluri JV, Rafil F, Sutherland JB, Wagner RD, Cerniglia CE (2001) Human health impact and regulatory issues involving antimicrobial resistance in the food animal production environment. Regul Res Perspect 1(1):10
31. Owino JO, Owido SFO, Chemelil MC (2006) Nutrients in runoff from a clay loam soil protected by narrow grass strips. Soil Tillage Res 88:116–122
32. Ramaswamy J, Prasher SO, Patel RM, Hussain SA, Barrington SF (2010) The effect of composting on the degradation of a veterinary pharmaceutical. Bioresour Technol 101:2294–2299
33. Sangster JL, Ali JM, Snow DD, Kolok AS, Bartelt-Hunt SL (2016) Bioavailability and fate of sediment-associated progesterone in aquatic systems. Environ Sci Technol 50(7):4027–4036
34. Sarmah AK, Meyer MT, Boxall ABA (2006) A global perspective on the use, sales, exposure pathways, occurrence, fate and effects of veterinary antibiotics (VAs) in the environment. Chemosphere 65(5):725–759
35. Sassman SA, Lee LS (2005) Sorption of three tetracyclines by several soils: assessing the role of pH and cation exchange. Environ Sci Technol 39(19):7452–7459
36. Song W, Guo M (2014) Residual veterinary pharmaceuticals in animal manures and their environmental behavior in soils. In: He Z, Zhang H (eds) Applied manure and nutrient chemistry for sustainable agriculture and environment
37. Soni B, Bartelt-Hunt SL, Snow DD, Gilley J, Marx D, Woodbury B, Li X (2015) Effect of narrow grass hedges on the transport of antimicrobial and antimicrobial resistance genes in runoff following land application of swine slurry. J Environ Qual 44(3):895–902
38. Stone JJ, Clay SA, Zhu Z, Wong KL, Porath LR, Spellman GM (2009) Effect of antimicrobial compounds tylosin and chlortetracycline during batch anaerobic swine manure digestion. Water Res 43:4740–4750
39. Szogi AA, Vanotti MB, Ro KS (2015) Methods for treatment of animal manures to reduce nutrient pollution prior to soil application. Curr Pollut Rep 1(1):47–56
40. Werner JJ, McNeill K, Arnold WA (2009) Photolysis of chlortetracycline on a clay surface. J Agric Food Chem 57(15):6932–6937
41. Zhang Y, Krysl RG, Ali JM, Snow DD, Bartelt-Hunt SL, Kolok AS (2015) Impact of sediment on agrichemical fate and bioavailability to adult female fathead minnows: a field study. Environ Sci Technol 49(15):9037–9047

Dr. Shannon L. Bartelt-Hunt received her B.S. in Environmental Engineering from Northwestern University in 1998 and her M.S. and Ph.D. in Civil Engineering, specializing in environmental engineering, from the University of Virginia in 2000 and 2004, respectively. She was a postdoctoral research associate in the Department of Civil, Construction, and Environmental Engineering at North Carolina State University from 2004 to 2005. In 2006, she joined the faculty in the Department of Civil Engineering at the University of Nebraska-Lincoln. At the University of Nebraska, she teaches courses on environmental engineering, engineering chemistry, and solid waste management. She holds a courtesy appointment in the Department of Environmental, Occupational, and Agricultural Health in the School of Public Health at the University of Nebraska Medical Center. Her research interests focus on the fate of biologically active contaminants in the environment, such as steroid hormones, pharmaceuticals, and the prion protein. She is specifically interested in evaluating the environmental impacts of agricultural production practices and water reuse in agriculture. She has authored over 95 peer-reviewed publications and book chapters and has served as an investigator or co-investigator on over $5 million in extramurally funded research. In 2012, she received a CAREER award from the National Science Foundation, and in 2015, she was recognized as part of a group receiving the Grand Prize for University Research from the American Academy of Environmental Engineers and Scientists. She is particularly interested in research at the intersection of environmental engineering, and human and animal health.

Shannon and her husband, George, have two children, Sam (age 12) and Alden (age 9). Outside of her work she enjoys reading, biking, and gardening, or at least she did prior to becoming a part-time "Uber driver" for her children and their many activities. Her interest in environmental research is grounded, in part, from growing up in a small farming community in Iowa. Her interest in science was motivated and encouraged by her wonderful high school biology and chemistry and physics teachers, Mr. Ralph O. Kaufman III and the late Mr. Elon (Lon) Rosine at Mediapolis High School. Her first research experiences were at Northwestern University under the guidance of Dr. Barbara-Ann Lewis and in a summer Research Experiences for Undergraduates (REU) program at Notre Dame University, directed by Dr. Stephen Silliman. Her summer REU experience confirmed to her that engineering is a helping profession. These experiences lit a passion for research that continued through her graduate study at the University of Virginia as well as during her postdoctoral training under Dr. Morton Barlaz at NC State University, and continue to this day.

Chapter 11
Understanding Soil-Contaminant Interactions: A Key to Improved Groundwater Quality

Maria Chrysochoou

Abstract Understanding the speciation of metal contaminants and their interactions with soils, sediments, and hazardous waste is critical both to predicting their mobility in the subsurface and to devising successful remediation approaches. Experimental-spectroscopic techniques and geochemical modeling can be coupled toward the study of metal interactions. This chapter will discuss contributions of infrared and X-ray-based techniques to study of the speciation of hexavalent chromium in two media: a Cr(VI)-contaminated soil from a plating facility and pure iron oxides—minerals that are abundant in natural soil environments, including the plating facility. The use of these techniques to inform treatment design and fate and transport models will be highlighted.

11.1 Chromium Presence and Mobility in the Environment

Chromium (Cr) is one of the most frequently detected metal contaminants in federal facilities, both in the Department of Energy (DoE) and Department of Defense (DoD) sites. It is also frequently found in industrial facilities, such as the metallurgic, tanning, and plating industries. As of 2012, Cr has been identified in 49% of the 1270 hazardous waste sites currently on the National Priorities List (NPL) under the federal Superfund program, making it second only to lead (Pb) as the most frequently cited metal pollutant. Moreover, less than 20% of those sites have been deleted from the NPL (i.e., cleaned up) since 1980. The state of New Jersey has a dedicated chromium cleanup program (Hudson County Chromate Project) listing 212 sites in that county alone, due to the widespread disposal of chromite ore processing residue (COPR) in the past decades. Chromium production has since shifted to countries such as China, India, and Pakistan, causing widespread pollution problems there [9, 22]. In addition to anthropogenic sources of chromium that cause soil

M. Chrysochoou (✉)
Department of Civil and Environmental Engineering, University of Connecticut,
Storrs, CT, USA
e-mail: maria.chrysochoou@uconn.edu

© Springer Nature Switzerland AG 2020
D. J. O'Bannon (ed.), *Women in Water Quality*, Women in Engineering
and Science, https://doi.org/10.1007/978-3-030-17819-2_11

and groundwater pollution, naturally occurring chromium in groundwater has also gained attention in recent years [6]. Soils and sediments in areas with ultramafic geologic background contain high concentrations of chromium, which may release in groundwater under the right geochemical conditions, and reach levels that exceed current standards for drinking water quality. Ultramafic rocks are characterized by high levels of magnesium and low levels of silica: while they are the main constituent of the earth's mantle, they are found only in certain regions near the surface [23].

The chemical and toxicological behavior of chromium is well-known, as are the processes that control its fate and transport in the environment. Figure 11.1 presents an overview of these processes, discussed in early studies [25].

Chromium has two major oxidation states in the environment—trivalent chromium (Cr(III)), which is an essential nutrient, and hexavalent chromium (Cr(VI)), which is a human carcinogen. In addition to the difference in toxicity, the two forms of chromium have substantially different geochemical behaviors. Cr(III) behaves similar to several other heavy metals (e.g., Pb, Ni) and is present as a cation (i.e., Cr^{3+}, $CrOH^{2+}$, $Cr(OH)_2^+$) in aqueous solution, while Cr(VI) is an oxyanion, present as CrO_4^{2-} or $HCrO_4^-$ in solution. Accordingly, the reactions that control their mobility in soil and groundwater are typical of the respective groups, cationic metals, and oxyanions. There are four types of reactions that influence metal behavior in the environment, including Cr [26]:

- *Aqueous complexation*: the formation of complexes with organic ligands or other ions such as chloride can increase the mobility of metal cations.
- *Precipitation*: cationic metals typically form insoluble hydroxides, including chromium and often also sulfides.

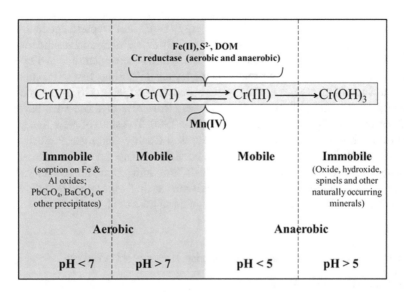

Fig. 11.1 Overview of chromium geochemistry

- *Sorption*: this refers to the formation of electrostatic or covalent bonds with reactive functional groups on mineral surfaces.
- *Redox reactions:* for redox-active elements such as Cr, these indirectly influence mobility by changing the predominant species that participate in sorption and precipitation reactions.

Nontoxic Cr(III) is typically insoluble at pH values greater than 5 due to the formation of insoluble oxyhydroxides as illustrated in Fig. 11.1. In ultramafic environments, Cr(III) is bound in even more insoluble minerals, such as chromite and serpentine [6]. Cr(III) mobility only increases in acidic environments and can also be enhanced by organic complexation [001].

Conversely, anionic Cr(VI) is mobile in aqueous environments, given that it forms very few precipitates, including $PbCrO_4$ and $BaCrO_4$. Other chromate salts such as chromatite ($CaCrO_4$) and commercial products such as $K_2Cr_2O_7$ have very highly solubility and are absent or infrequent in the natural environment [6]. Precipitation is typically prevalent in alkaline, cementitious matrices such as COPR and cements, in which Cr(VI) forms phases such as chromate ettringite and chromate hydrocalumite [2]. Sorption to positively charged mineral surfaces such as iron and aluminum oxyhydroxides is a potential retardation mechanism for Cr(VI) fate and transport in soil and groundwater [11].

The most frequent remediation approach for both Cr(VI)-contaminated soils and other waste including COPR is the reduction to immobile Cr(III). This can occur using various types of reductants, including ferrous and zerovalent iron, reduced forms of sulfur, and microorganisms. The success of a reductive treatment depends on several factors, including pH, presence of competing reactants, and availability of Cr(VI) in solution. Redox reactions are efficient and rapid in solution: if Cr(VI) is precipitated or entrapped in the solid phase, reduction may not occur quickly. A prerequisite for irreversible reduction is the absence of high-valence manganese oxides such as birnessite, which can facilitate Cr(III) reoxidation. Thus, the knowledge of the particular form of Cr(VI) present in the solid, and whether it is a contaminated soil or waste, is necessary for optimal treatment design.

A wide variety of analytical, spectroscopic, and modeling approaches is available in the literature to study the speciation of metals. This chapter focuses on two techniques that have been widely applied to study the interaction of various metals with minerals, soils, and waste: micro-X-ray fluorescence spectroscopy (μXRF) and Fourier transform infrared (FTIR) spectroscopy.

11.2   X-Ray-Based Analysis of Cr(VI) in Contaminated Soil

Chromium plating facilities have frequent occurrences of Cr(VI) contamination in soil and groundwater, as disposal practices in the past were conducive to releases of liquid processing waste as well as sludges in the immediate surrounding. In a typical example at a site in northeastern Connecticut, drippings from the Cr plating process and wastewater were directly discharged into an adjacent wetland for decades until environmental regulations took effect. The facility is still operational, and a pump-and-treat treatment system for contaminated groundwater has operated for approximately 20 years now. Because pump and treat does not address the source of chromium, a study on different in situ reduction alternatives took place. The studies examined reductants such as calcium polysulfide and nanoscale zerovalent iron [4, 5].

Treatability studies are a typical element of a remedial investigation, even for chemicals that have been proven to be effective. The necessary dosage to achieve the desired treatment goals in heterogeneous soil and groundwater matrices is best determined by conducting short-term batch studies, in which the soil and/or groundwater are mixed with variable amounts of reductant and the contaminant is monitored to determine treatment success. The necessary dosage for in situ treatment is almost always higher compared to the stoichiometry predicted by the reaction, given that competing reactants are present in natural environments. For example, ferrous iron is a common reductant for Cr(VI) that can be oxidized by oxygen or precipitate as a hydroxide. Treatability studies do not typically take into account individual competing reactions, but they empirically determine the total amount of reactant necessary under a specific set of conditions.

Chrysochoou et al. [4] presented the results of a treatability study that sought to determine the minimum amount of calcium polysulfide (CaS_5) required to treat contaminated soil from the Connecticut facility, both below and above the groundwater table. Initially, traditional analytical tests such as alkaline digestion with colorimetric analysis (U.S. Environmental Protection Agency (EPA) methods 3060 and 7196) and the Synthetic Precipitation Leaching Procedure (SPLP) were used to monitor treatment success. Two major conclusions were reached from these analyses:

- EPA 3060A caused treatment artifacts by reducing Cr(VI) during the test, an observation that was previously established with chromite ore processing residue [8].
- The reduction proceeded very slowly, with continuous decrease in the SPLP observed even after 1 year of incubation.

The reaction kinetics of aqueous chromate with calcium polysulfide are relatively rapid [3] and did not justify the second observation: only a slow leaching of chromate of the soil could account for such slow kinetics. To determine the mechanism that could account for the slow leaching of what is known as a highly mobile element, spectroscopic analyses were conducted at beamline 10.3.2 at the Advanced Light Source (ALS). The micro-X-ray fluorescence (μXRF) beamline allowed the mapping of the elemental composition of the soil with a resolution of 10 by 10 μm^2. Soil samples were prepared as thin sections, which insure both a perfectly flat surface, so that the cross section of soil grains can be studied. In addition, a technique was developed at the beamline to isolate the signal of hexavalent chromium from total chromium, which allows the detection of Cr(VI)-binding species within the soil matrix.

Figure 11.2 shows an example of the μXRF maps generated by XRF analyses. Each pixel of the image, sized $10 \times 10\ \mu m^2$, contains the XRF signal for a variety of elements, which can be illustrated in tricolor maps for any desired combination. The color intensity represents the relative concentration of the element with respect to the maximum X-ray intensity recorded on the map, and thus the relative distribution of elements and their associations can be determined both by visual inspection and by a statistical analysis of the intensities recorded over all pixels. The maps shown in Fig. 11.2 are 3000 by 1700 μm^2 in size. Individual soil grains can be easily identified in Fig. 11.2a that depicts the abundant elements K, Si, and Fe, including angular quartz particles and K-feldspars. Figure 11.2b depicts Cr(VI), Cr(III), and Pb and yields the following observations:

- Cr(III) either coexists with Fe in large grains or is precipitated on the surface of the larger feldspar particles as what appears to be a thin coating.
- Cr(VI) is mostly intimately associated with Pb, in grains that range in size from a few microns to a large particle of ~300 μm in length. One large grain contains high Cr(VI) concentrations that is a Fe-coated silica grain (Fig. 11.3).

The observation of Cr(VI) association with Pb was especially surprising, given that the soil was never previously analyzed for the presence of Pb, a metal that was

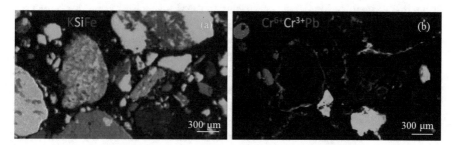

Fig. 11.2   Tricolor micro-XRF map of Cr(VI)-contaminated soil from a plating facility in CT (**a**) potassium in red, silicon in green, iron in blue color; (**b**) hexavalent chromium in red, trivalent chromium in green, lead in blue color

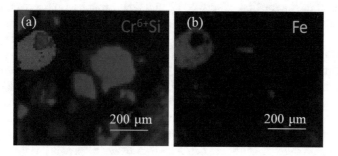

Fig. 11.3   Micro-XRF maps of high Cr(VI) grain from Fig. 11.2 (**a**) hexavalent chromium in red, silicon in blue; (**b**) iron in green

not explicitly part of the chromium plating process. The source of Pb is still not exactly known; however, it was confirmed that it was present in the soil at a concentration of 15,000 mg/kg [4]. The total concentrations of Cr(VI) and Pb in the soil indicated that up to 50% of the Cr(VI) could be present as $PbCrO_4$, which is one of the few highly insoluble Cr(VI) precipitates. The remaining Cr(VI) was observed to be associated with Fe, as Cr(VI) binding on the surface of iron oxide minerals at the soil pH (5.5–6) is a well-known retardation mechanism [11]. These observations explained the behavior of the soil upon treatment: a large fraction of Cr(VI) was bound in $PbCrO_4$ that is difficult to dissolve; therefore the reductive treatment proceeded very slowly. In such a scenario, optimization of reductive treatment requires persistent reductant or repetitive small reductant doses over longer time frames.

11.3 Spectroscopic Studies and Modeling of Cr(VI) Interactions with Iron Oxides

Modeling of metal sorption onto soil surfaces has traditionally been incorporated into the empirical distribution factor (K_d). However, this approach is extremely limited, because it cannot account for changes in groundwater chemistry (pH, ionic strength, competitive ions) and is restricted to the site-specific system for which the empirical K_d value was obtained. In the last 20 years, a significant effort has been made to shift to a more mechanistic description of sorption reactions through the development of surface complexation models (SCMs). SCMs provide a thermodynamic description of the reaction between soil surface ligands and sorbed compounds, taking into account the charge associated with each. Major advantages of SCMs include the ability to incorporate these reactions in software codes modeling aqueous chemistry and the ability to model sorption reactions across geochemical gradients [13]. In sorptive exchange, the effort associated with obtaining the necessary parameters to adequately describe SCMs increases substantially. A large number of required parameters are needed for the mechanistic level of the model, and it is difficult to obtain all these parameters.

All SCM formulations rely on the same principles: mineral surfaces are charged as a result of the interaction of surface H^+ and OH^- groups with the overlying solution. The surface charge is balanced through the formation of electrostatic or covalent bonds between surface groups and dissolved cations or anions. The various models (Constant Capacitance Model (CCM) [28], Double Layer Model (DLM) [10], Triple Layer Model (TLM) [7], and Charge Distribution-Multisite Complexation (CD-MUSIC) model [15]) differ in the way they formulate surface group reactions, the types of bonds they consider, and the mathematical description of the electrical field near the charged surface. The assumptions of each model influence its flexibility in describing sorption reactions under various conditions and the number of parameters required to calibrate it. A heavy emphasis has been placed in the refinement of SCMs in the last few years, utilizing complementary experimental and computational approaches to identify the types of surface complexes that a compound forms under different geochemical conditions. Various types of spectroscopy combined with predictions of molecular modeling calculations can be used to narrow down the surface complexation reactions and constrain the number of parameters needed to adequately describe adsorption, as well as validate the model predictions.

The strongest sorbents for Cr(VI) are iron oxides, which have various forms in natural soils, such as ferrihydrite ($Fe_2O_3 \cdot 0.5H_2O$), goethite (α-FeOOH), and hematite (α-Fe_2O_3) [26]. Chromate (CrO_4^{2-}) can form three types of complexes on these surfaces, as illustrated in Fig. 11.4: outer sphere, which only relies on electrostatic attraction between the positively charged surface and the anion and inner sphere, in which the ion forms a covalent bond directly with the reactive oxygen groups present on the surface [27]. These complexes may also be protonated or unprotonated, yielding a total of six possible surface species. The surface complexation reaction depends on the type of complex, and the electrostatic gradient near the surface is also influenced. Ultimately, the species utilized to describe the surface reaction affect the parametrization and, hence, the sensitivity of the model to varying conditions. Specifically for chromate on iron oxides, the CD-MUSIC model uses (where $>FeOH^{-1/2}$ denotes the reactive surface site):

$$> FeOH^{-1/2} + H^+ + CrO_4^{2-} \rightarrow > FeOCrO_3^{-1/2} + H_2O \qquad (11.1)$$

$$2 > FeOH^{-1/2} + 2H^+ + CrO_4^{2-} \rightarrow > \left(FeO\right)_2 CrO_2^- + 2H_2O \qquad (11.2)$$

The first reaction corresponds to the formation of a monodentate surface complex, as shown in Fig. 11.2, while the second reaction corresponds to the formation of a bidentate surface complex. Each reaction has its own corresponding log K and

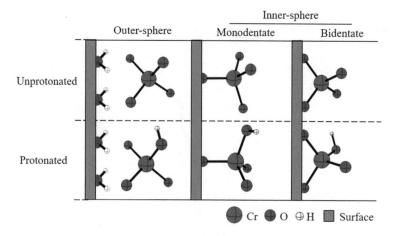

Fig. 11.4 Complexation modes of CrO_4^{2-} on iron oxide surfaces and modeling studies assuming a specific complex type

mass action coefficient, as the reaction stoichiometries for the reactants are different. Surface complexation models adopted several combinations of the possible species, by determining log K_s from fitted macroscopic sorption data. Most early models, i.e., prior to 1990, had no corroborating information on the type of complexes formed. Later on, several studies utilized spectroscopic techniques to study chromate complexation on iron oxides. Specifically, three techniques were implemented, mostly to study chromate sorption on goethite [1, 12, 16]: extended X-ray absorption Fourier transform spectroscopy (EXAFS), transmission and attenuated total reflectance (ATR) FTIR, and X-ray photoelectron spectroscopy. It was later found that techniques that require the use of dry samples skew the results toward the identification of inner sphere complexes and are thus less reliable to accurately determine surface speciation. Conversely, both EXAFS and ATR can test samples in equilibrium with a solution to obtain a spectroscopic signal that results from the interaction of the ligand with the surface without interference from the solvent (water).

Figure 11.5 illustrates the use of ATR spectra to differentiate between aqueous and sorbed chromate species. Aqueous chromate at pH values >6.5 has a single peak at 880 cm^{-1}. Aqueous bichromate (HCrO$_4^-$) has a predominant peak at 950 cm^{-1} and a smaller peak at 880 cm^{-1}. When chromate coordinates on a mineral surface, the symmetry of the molecule is reduced, and the peaks split into several overlapping peaks. Resolving the number and location of these peaks can be performed through several techniques, such as the use of difference spectra, inverse second derivative spectra, and multivariate curve resolution analysis [24].

These techniques were utilized to study chromate interaction with ferrihydrite [17, 21], hematite [18, 21], boehmite [19], and Al-bearing ferrihydrite [20]. Collectively, these studies yielded the following conclusions:

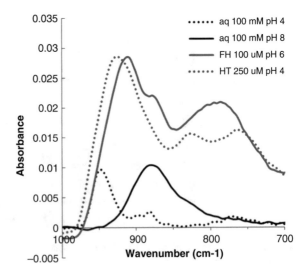

Fig. 11.5 Comparison of aqueous chromate (CrO$_4^{2-}$) and bichromate (HCrO$_4^-$) spectra at pH 8 and 4, respectively, with chromate sorbed on ferrihydrite (FH) and hematite (HT)

- Chromate forms the same types of complexes on all three iron oxides, goethite, ferrihydrite, and hematite, i.e., an unprotonated monodentate and an unprotonated bidentate complex.
- The bidentate complex is generally dominant, especially at higher surface coverage and pH values below 7.
- Aluminum oxides and incorporation of Al in iron oxides favor the formation of weaker outer-sphere complexes that preferentially displace monodentate complexes from the surface.

Observations indicate that the hydrogen is not associated with the chromate molecule regardless of the aqueous pH. The latter observation was established by performing FTIR experiments in H_2O and D_2O. Replacement of a hydrogen atom with a heavier deuterium atom in a molecule would cause a shift in the associated vibrations. Since no shift in peaks was observed, it follows that the chromate surface complex is not associated with a hydrogen species [17]. This observation allows for the elimination of protonated surface complexes shown in Fig. 11.2 for consideration in the SCM.

These observations indicate the chromate sorption on different types of iron oxides can be described using the same reactions (11.1) and (11.2) that only the presence of aluminum imposes the need for a third reaction of the outer-sphere complex. More reactions require more log K_s have to be determined, and macroscopic batch sorption tests do not supply the information necessary to determine such unique constants. This is a common conundrum with modeling exercises: the model fit improves when more parameters are considered; however the datasets are often insufficient.

Spectroscopy can further support the modeling problem by providing quantitative estimates for the distribution of surface species with variables such as pH and surface coverage. This approach is still at its infancy, with only a limited number of studies including quantitative analysis of ATR [18] and EXAFS spectra [14]. Its wider adoption can substantially aid the constraining of log K values for individual surface complexes, which cannot be determined using traditional batch isotherms. A large emphasis has been placed on the qualitative investigation of the mechanisms of surface complexation; however proper constraints for the resulting surface complexation models are still needed.

11.4   Summary and Outlook

Spectroscopic techniques are valuable for scientific research and can also be used to aid the decision-making process for environmental contamination and remediation. Specifically, spectroscopy can shed light into:

- The behavior and timing of contaminant release from the solid into the liquid phase; this information can be used to optimize the design and execution of treatability studies for contaminated soils.

- The long-term stability of precipitates that immobilize contaminants; similarly, this information can be used to support Remedial Action Plans that utilize in situ immobilization as the treatment approach for soil and/or groundwater.
- Secondary reactions that may interfere with in situ remediation processes (fouling, clogging, secondary mobilization, potential for long-term redox reactions). The identification of precipitates using spectroscopic techniques can point to the mechanism that caused the secondary reactions, which is not always apparent using more traditional analytical data.
- Mechanisms for reactive transport modeling. The evolution of traditional models that use the K_d (distribution factor) approach can be informed by spectroscopy. There is a wide gap between the state of the art and the state of the practice in this field, and there is substantial development necessary to bridge that gap.

A key challenge for environmental researchers is to communicate these capabilities to the industry and establish collaborations and offer services that are accessible and affordable for more widespread use of spectroscopic techniques.

References

1. Abdel-Samad H, Watson PR (1997) An XPS study of the adsorption of chromate on goethite (α-FeOOH). Appl Surf Sci 108:371–377
2. Chrysochoou M, Dermatas D (2006) Evaluation of ettringite and hydrocalumite formation for heavy metal immobilization: literature review and experimental study. J Hazard Mater 136(1):20–33
3. Chrysochoou M, Ting A (2011) A kinetic study of Cr(VI) reduction by calcium polysulfide. Sci Total Environ 409:4072–4077
4. Chrysochoou M, Ferreira D, Johnston C (2010) Calcium polysulfide treatment of Cr contaminated soil. J Hazard Mater 179:650–657
5. Chrysochoou M, Johnston C, Dahal G (2012) A comparative evaluation of Cr(VI) treatment in contaminated soil by calcium polysulfide and nanoscale zero valent iron. J Hazard Mater 201–202:33–42
6. Chrysochoou M, Theologou E, Bompoti N, Dermatas D, Panagiotakis I (2016) Occurrence, origin and transformation processes of geogenic chromium in soils and sediments. Curr Pollut Rep 2(4):224–235
7. Davis JA, James RO, Leckie JO (1978) Surface ionization and complexation at the oxide/water interface. J Colloid Interface Sci 63:480–499
8. Dermatas D, Chrysochoou M, Moon DH, Grubb DG, Wazne M, Christodoulatos C (2006) Ettringite-induced heave in chromite ore processing residue (COPR) upon ferrous sulfate treatment. Environ Sci Technol 40(18):5786–5792
9. Du J, Lu J, Wu Q, Jing C (2012) Reduction and immobilization of chromate in chromite ore processing residue with nanoscale zero-valent iron. J Hazard Mater 215–216:152–158
10. Dzombak DA, Morel FMM (1990) Surface complexation modeling: hydrous ferric oxide. John Wiley & Sons, New York
11. Fendorf SE (1995) Surface reactions of chromium in soils and waters. Geoderma 67:55–71
12. Fendorf S, Eick MJ, Grossl P, Sparks DL (1997) Arsenate and chromate retention mechanisms on goethite. 1. Surface structure. Environ Sci Technol 31:315–320
13. Goldberg S, Criscenti LJ, Turner DR, Davis JA, Cantrell JK (2007) Adsorption-desorption processes in subsurface reactive transport modeling. Vadose Zone J 6:407–435

14. Gu C, Wang Z, Kubicki JD, Wang X, Zhu M (2016) X-ray absorption spectroscopic quantification and speciation modeling of sulfate adsorption on ferrihydrite surfaces. Environ Sci Technol 50(15):8067–8076
15. Gustafsson JP, Persson I, Oromieh AG, van Schaik JW, Sjöstedt C, Kleja DB (2014) Chromium(III) complexation to natural organic matter: mechanisms and modeling. Environ Sci Technol 48(3):1753–1761. https://doi.org/10.1021/es404557e. Epub 2014 Jan 22
16. Hiemstra T, Van Riemsdijk W (1996) A surface structural approach to ion adsorption: the charge distribution (CD) model. J Colloid Interface Sci 179:488–508
17. Hsia TH, Lo SL, Lin CF, Lee DY (1993) Chemical and spectroscopic evidence for specific adsorption of chromate on hydrous iron oxide. Chemosphere 26:1897–1904
18. Johnston CP, Chrysochoou M (2012) Investigation of chromate coordination on ferrihydrite by in situ ATR-FTIR spectroscopy and theoretical frequency calculations. Environ Sci Technol 46(11):5851–5858
19. Johnston CP, Chrysochoou M (2014) Mechanisms of chromate adsorption on hematite. Geochim Cosmochim Acta 138:146–157
20. Johnston CP, Chrysochoou M (2015) Mechanisms of chromate adsorption on boehmite. J Hazard Mater 281:56–63
21. Johnston C, Chrysochoou M (2016) Mechanisms of chromate, selenate, and sulfate adsorption on Al-substituted ferrihydrite: implications for ferrihydrite surface structure and reactivity. Environ Sci Technol 50(7):3589–3596
22. Kabengi N, Chrysochoou M, Bompoti N, Kubicki J (2017) An integrated flow microcalorimetry, infrared spectroscopy and density functional theory approach to the study of chromate complexation on hematite and ferrihydrite. Chem Geol 464:23–33
23. Matern K, Kletti H, Mansfeld T (2016) Chemical and mineralogical characterization of chromite ore processing residue from two recent Indian disposal sites. J Hazard Mater 155:188–195
24. Oze C, Fendorf S, Bird KD, Coleman GR (2004) Chromium geochemistry of serpentine soils. Int Geol Rev 46:97–126
25. Peak D, Elzinga EJ, Sparks DL (2001) Understanding sulfate adsorption mechanisms on iron(III) oxides and hydroxides: results from ATR-FTIR spectroscopy. In: Selim HM, Sparks DL (eds) Heavy metals release in soils. Lewis Publishers, Boca Raton
26. Rai D, Eary LE, Zachara JM (1989) Environmental chemistry of chromium. Sci Total Environ 86:15–23
27. Sparks DL (2003) Environmental soil chemistry. Elsevier Academic Press, San Diego
28. Sposito G (1989) The chemistry of soils. Oxford University Press, New York
29. Stumm W, Kummert R, Sigg LM (1980) A ligand exchange model for the adsorption of inorganic and organic ligands at hydrous oxide interfaces. Croat Chem Acta 53:291–312

Maria Chrysochoou is a Professor and Head of the Department of Civil and Environmental Engineering at the University of Connecticut. She obtained her bachelor's degree in Physics at the Aristotle University of Thessaloniki in Greece, with a focus on environmental and atmospheric physics. She continued her studies at Technische Universität Dresden in Germany and obtained an M.Sc. in Environmental Engineering with focus on waste management and contaminated site remediation. She then moved to the USA to work on her Ph.D. in Environmental Engineering at Stevens Institute of Technology, researching the properties and treatment of chromite ore processing residue (COPR), a hazardous waste found in numerous sites across the Northeast USA. Dr. Chrysochoou also completed her postdoctoral studies at Stevens Institute of Technology, working on multiple projects involving COPR, lead (Pb) in US Army firing ranges, and soil stabilization using industrial by-products. She was hired as Assistant Professor at the University of Connecticut in 2007, promoted to Associate Professor in 2013 and Full Professor in 2019, and served as Program Director for Environmental Engineering from 2015 to 2018.

Dr. Chrysochoou's general research area is environmental geochemistry, with a focus on utilizing spectroscopic techniques to investigate speciation of metals on soils, mineral surfaces, industrial waste, and construction materials. Reuse and recycling of industrial by-products has been another topic of interest, and Dr. Chrysochoou and her team received the US Environmental Protection Agency People Prosperity and Planet award, in 2012, for their work on using local by-products to stabilize soils against erosion in Nicaragua. In 2013, she obtained a Marie Curie fellowship from the European Union and spent 2 years in Greece utilizing spectroscopy and other techniques to investigate the speciation and mobility of naturally occurring chromium in Greek aquifers that endangers local water supplies. Dr. Chrysochoou routinely teaches soil mechanics, site remediation, and geoenvironmental engineering and oversees senior design and thesis projects, involving undergraduate students in her research lab. She is a member of the American Society of Civil Engineers (ASCE) Geoenvironmental committee and has developed and offered ASCE webinars on geochemistry applications in civil engineering. Dr. Chrysochoou also serves as a subject matter expert for the Federal Highway Administration, consulting state departments of transportation on the use of X-ray fluorescence and infrared spectroscopy for quality assurance of common construction materials.

Dr. Chrysochoou grew up in Greece, where she enjoys spending summers by the sea with her daughter Artemis. Her personal goals are to finish reading the *New York Times* 100 Notable Books of the Year list at least once and attend the playoffs with her local women's volleyball team.

Notes on Some Sulpharsenites and Sulphantimonites from Colorado

by Ellen H. Swallow
B.S. thesis
1873
Department of Chemistry

In the summer of 1872, Professor Richards obtained specimens of the silver-bearing minerals of some of the Colorado mines.

They were called, by the mines, brittle silver or gray copper, but, as they had never been analyzed, nothing definite was known as to their composition.

In March 1873 I was requested to test these specimens sufficiently to ascertain whether they belonged under stephanite or tetrahedrite and whether the minerals from the different mines were of the same composition.

A qualitative analysis showed that all contained copper and in varying proportions.

These indications together with the results of the blowpipe tests divided the specimens into two groups.

Those from the Illinois and Walker mines contained much copper and little silver and both arsenic and antimony, while those from the Hercules and Terrible mines contained a very large amount of silver with comparatively little copper and antimony with traces only of arsenic.

The Illinois Lode, Central City, Colorado, yields both gold and silver by the dry assay although neither appear in the results given of the analysis by the wet way.

Luster, metallic. Color, dark gray. Streak, brown. Not easily fused.

Iron pyrite was abundantly distributed throughout the mass, and although the coarse powder was carefully examined under the microscope, some pyrite and much quartz evidently escaped separation.

This would give a slightly different ratio between (Ag. Cu. Fe) and (As. Sb. S.) from that which the exact analysis would give.

© Springer Nature Switzerland AG 2020
D. J. O'Bannon (ed.), *Women in Water Quality*, Women in Engineering and Science, https://doi.org/10.1007/978-3-030-17819-2

The ratio of tetrahedrite is 1:24, while the ratio of the percentages as given is about 1:2.

The mineral would therefore seem to be a true tetrahedrite.

The specimen from the Walker Mine, Montezuma, Colorado, was massive, in quartz rock, without galena.

Luster, metallic. Color, dark brown. Streak, red. Fusible.

The percentages given below of the Walker Mine are the results of a single analysis but suffice to indicate that the mineral is a tetrahedrite.

	Illinois		Walker	
SiO_2	15.04		0	
Pb	0		2.64	
Ag	Undetermined		5.91	
Au	Undetermined		–	
Cu	41.32		36.07	
Fe	6.24		.80	
Zn	6.00		0	
S	24.03		26.41	
Sb	1.62		Undetermined	71.83
As	6.31		Undetermined	
		100.56		

There were four specimens from Hercules Mine and five from Terrible Mine, both at Georgetown, Colorado.

The gangue rock from both mines seemed to be a tolerably fine granite containing very little mica.

The silver ore occurred, intermingled with finely crystallized galena and zine blende, in a vein or fissure lined on both sides with splendid crystals of quartz. It seemed, in some instances, to have been deposited after the galena and blende for it filled up the crevices in them and occupied the spaces between the quartz crystals.

In one case, it seemed almost a vein between the galena and blende.

Luster, metallic. Color, steel gray. Streak, black. Very brittle

The specimens were mainly massive although in two, something like crystalline form was noticed with striae along the length of the prisms.

On attempting to get them out separately for measurement, they broke into several pieces which, under a magnifying glass, seemed to present a micaceous, laminated structure. Thin scales were placed under a microscope, and the forms obtained by examination of the different specimens are roughly depicted below.

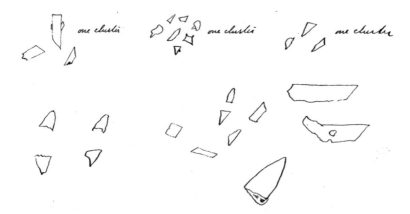

There was definiteness of form sufficient to indicate a crystalline structure, and a similarity of shape indicated that the crystalline form would be the same.

The specimens from Terrible Mine numbered 3, 4 and 5 with Hercules 2 and 4 were more silvery in luster and less massive in structure, rather more brittle, yet as the galena and blende were also in more perfect crystals, the outward appearance did not show a difference sufficiently marked to suggest a difference of composition.

The ore was in so small quantity and so intimately mingled with the other ores that it was very difficult to get a clear sample; those, from which the results tabulated below were obtained, were chipped off with a penknife and carefully examined with a microscope that all foreign minerals might be separated—still, only two were free from galena.

The object originally in view, in undertaking the investigation, was merely to see if the ore was of a uniform composition, and absolute accuracy is not claimed for any of the analyses, but they are reliable enough to base a theory upon.

Some analyses from Dana's Mineralogy are given for comparison.

	Hercules			Terrible			Stylo-typite	Poly-basite	Steph-anite	Tetrahedrite (freibergite)
	No. 1.	No. 2	No. 3	No. 1	No. 2	No. 3				
SiO_2	2.24	2.00	4.86	2.60	1.00	.80				
Pb	1.67	0	5.00	2.19	11.67	0				
					32.50					
Ag	36.67	60.06	44.68	47.00	28.00	57.76	8.30	68.55	68.51	31.29
Cu	18.04	10.18	12.48	9.65	29.20	13.35	28.00	3.36	.64	14.81
Fe	1.10	1.19	5.52	.57	1.90	0	7.00	.14	0	5.98
										Zn .99
Sb	17.80	Undet	Undet	Undet	Undet	Undet	30.53	11.53	14.68	24.63
As	Traces?	Traces?	Traces?	–	–	–				
S	22.70	17.42	17.45	18.90	19.00	20.33	24.30	15.55	16.42	21.17
Ratio	$1{:}2\frac{2}{3}$	$1{:}1\frac{4}{5}$	$1{:}1\frac{4}{5}$	$1{:}2\frac{1}{2}$		$1{:}2$	$1{:}2\frac{2}{3}$	$1{:}1\frac{1}{2}$	$1{:}2$	$1{:}2\frac{1}{4}$

It will be seen that the percentages do not quite agree with any given analysis. Hercules No. 1 and Terrible No. 1 are probably tetrahedrite although containing more silver than any previously analyzed specimen.

These two were darker in color and more compact than the others. Terrible No. 2 doubtless belongs with them, but the amount of galena which was evidently mingled with the sample throws it out of account.

Terrible No. 3 has the ratio of stephanite, while Hercules No. 2 and No. 3 are between polybasite and stephanite. As the antimony was estimated by difference, a more accurate analysis might change the ratios so that they would agree with stephanite, but from the amount of copper contained in all the samples, it would seem that they were nearer to polybasite.

Apparatus suggested by Professor Crafts for the analysis of tetrahedrites contained possibly Cu, Ag, Fe, Zn, Hg, Sb, As, and S and also as impurities Pb S, Zn S, Cu S, and SiO_2.

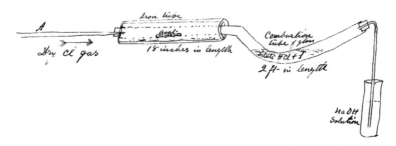

The small straight tube A is long enough to allow the sliding back of the iron tube quite off from the glass tube.

After the weighted substance is placed in the tube (in a porcelain boat), dry chlorine gas is passed over it without heating as long as there appears to be any action. Then the iron tube is pushed forward over the combustion tube and heated to dull redness by an ordinary combustion furnace.

As to the materials employed, most of the results given in the table were obtained by the chlorine method, the residue from treatment with chlorine being first treated, in the boat, with strong, hot nitric acid; then well washed with hot water, the residue thus obtained was separated by decantation or by filtration if necessary, treated with ammonia which dissolved the chloride of silver and left only silica which was filtered off, dried, and weighed. The silver was reprecipitated by nitric acid, weight as chloride. The copper was precipitated from the first nitric acid solution as subsulfide [sic] and the iron determined in the filtrate.

The solutions containing the volatile chlorides were used in making experiments on the method. The lead was obtained from the evaporation with sulfuric acid for the precipitation of copper as subsulfide.

Duplicate determinations were made as follows. The sulfides were oxidized with nitric acid, evaporated and taken up with nitric acid and water, filtered, the silver precipitated by dilute chlorhydric acid, the filtrate evaporated with sulfuric acid with the addition of tartaric and sulfurous acids. The lead was filtered out, the solution heated and then treated with hyposulfite of soda which precipitates copper arsenic and antimony as sulfides, the filtrate is oxidized, evaporated to a small bulk, the separated sulfur filtered out, and the iron and zinc are determined as usual. The precipitate is treated with sulfide of sodium, washed, dried, and ignited with sulfur. The solution in sulfide of sodium is acidified with chlorhydric acid, the precipitated sulfides of arsenic and antimony filtered out and oxidized and separated and determined by Fresenius – (? 155).[1]

In some cases this method is very convenient, not requiring the apparatus of the chlorine method and giving the arsenic and antimony in a more convenient and accurate form.

The arsenic and antimony are completely precipitated with the copper if the solution is dilute enough and kept at just the boiling point for a few minutes.

In analyzing these minerals, I was puzzled to separate the copper in the presence of arsenic and antimony, and the battery was hardly practicable. I had just been estimating copper in some other minerals, not containing arsenic or antimony, as subsulfide according to Fresenius.

As I found it a quick process, I wished to try to use it in the case of the minerals under examination, but I could find no reference to the behavior of arsenic or antimony compounds with sodium hydrosulfide, and no one could tell me what the reaction would be. I did not know whether I should find the arsenic and antimony in the precipitate with the copper or in the filtrate with the zinc or in both places.

In order to settle the question, known solutions of arsenic and antimony were treated at first separately and afterward together with a solution of sodium hydrosulfite and the conditions of complete precipitation determined as above.

At the time this was an entirely original investigation (March 1873). I have recently (May 1873) noticed an account of the preparation of sulfides of arsenic and antimony for calico printing and for some other technical uses by means of hyposulfite Chemie Appliqe'e (Barreswil) 1859[2]; Graham-Otto Band.[3]

I have yet seen no mention of it as a method of separation or estimation.

I propose to try to use hyposulfite for determining either arsenic or antimony or both together and in that case separating and estimating the arsenic, volumetrically by an apparatus something like that figured below; the sulfurous acid which is evolved may be received into a flask, and after cooling, the contents of both flasks may be titrated with iodine.

---

[1] Fresenius, C. Remigius. 1866. Anleitung zur quantitativen chemischen Analyse : oder die Lehre von der Gewichtsbestimmung und Scheidung der in der Pharmacie, den Künsten, Gewerben und der Landwirthschaft häufiger vorkommenden Körper in einfachen und zusammengesetzten Verbindungen, für Anfänger und Geübtere. Braunschweig: Friedrich Vieweg und Sohn.

[2] Appliqee: Repertoire de chimie appliquee, 1859.

[3] Graham-Otto: Lehrbuch der physikalischen und theoretischen Chemie (Graham-Otto's Ausführliches lehrbuch der Chemie, Band 1.) 1857. Buff, Johann Heinrich; Kopp, Hermann; Zamminer, Friedrich Georg Karl. Braunschweig : Friedrich Vieweg und Sohn.

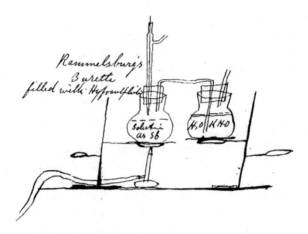

There must be no nitric acid and a very slight excess of chlorhydric acid.

I have begun a series of investigations on the behavior of the other metals with hyposulfite with a view to employing hyposulfite in qualitative analysis, in some instances at least, instead of sulfidric acid.

I think it might be used as a preliminary test and save much time - - -

Ellen H. Swallow

Note: Douglas and Prescott since published (illegible)

Index

© Springer Nature Switzerland AG 2020
D. J. O'Bannon (ed.), *Women in Water Quality*, Women in Engineering
and Science, https://doi.org/10.1007/978-3-030-17819-2

Printed in the United States
By Bookmasters